a

tiny story

a

tiny

story

by

fernando maquieira

contents

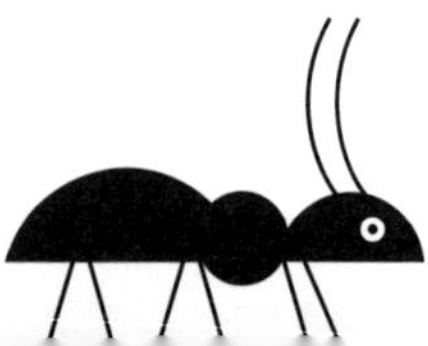

1 Lea the ant

14

2 Mission Pavarotti

18

3 The far side of the garden

28

4 Alfi

38

5 The Abandoned

48

6 The battle

64

for you

"The ant who made the find had no idea how lucky she was."

Juan José Arreola, *The marvellous milligram*

We humans use the word *ant* so we can understand each other and know who we are talking about. Ants, in their own language, go by a name that is impossible for us Chatterboxes, as they like to call us, to write or pronounce. They don't care much for us Chatterboxes because we often step on them by accident, or not – which is why they always run for cover when they see us coming. They have loads of ways of communicating, but their favourite way is by using their antennae, which they rub against one another to talk about their lives. As with this tiny story, passed from antennae to antennae by the dwellers of the nameless garden, a story that I picked up from someone who heard it one night from singing crickets.

1

Lea the ant

Lea the ant emerges at dawn for the first time from the anthill where she was born just three months ago. The opening to the anthill is a crack in the base of a wall around a farmhouse, a wide hole shaped like a donkey's yawn. In the morning, the light just about slants inside the hall, which opens out into a maze of dark tunnels, like wide roads carved out by ants, a huge underground city, daily teeming with the ants that live there in their thousands, bustling about among their sisters, since they are all daughters of the same queen. The gate is watched day and night by the guards, large and very serious ants who control the entrance, keeping a strict note of every milligram of food coming in and of who brings it.

As Lea leaves the dark safety of the anthill, she gets a bit scared, although she is good at hiding it; it's as if she doesn't mind the glare and the many strange smells wafting through the air of the outside world. All she's known up until now of the outside world were the adventures the veterans in the nest told her about since she was only larva. She was always asking them to retell their stories over and over. So Lea knows very little about the new life she's embarking on. Everything that runs, flies and grows above the anthill holds mystery for her.

Briskly striding along, she keeps in line behind her older sisters. She's the youngest of a small group from the ant colony out on a mission to find cherries, fresh lettuce leaves or any other kind of food to bring back to the anthill. It's like shopping without needing money. They're walking around with their heads held high, their antennae twitching every which way, constantly tracking everything with their built-in radar – a really useful skill that allows them to decode signals hidden about the garden or read the secret messages left by the scouts, but above all to sniff out the trail of food that's ready for eating.

2

Pavarotti Mission

The squad has stopped along the usual path for workers, to talk to a group of female scouts on their way home from their daily rounds. Chatting by antennae, they tell them to follow the scent trail they have left, like a treasure map, from the root of the dried fig tree to the cage of Pavarotti, the house canary, where there is a decent pile of birdseed on the ground, easy to collect.

"Just be careful," warns the scout leader. The Chatterboxes have moved the cage, and you have to be really careful not to cross paths with the ants from the far side of the garden, because the path to the birdseed runs through their patch, and you know how unfriendly they are. So, you'll have to be quick. Listen very closely with your antennae and make sure you get there without being seen."

"That's quite a mission for your first day!" says Lea's pal alongside her. "But don't you worry: if you stick with me and do as I say, you'll have nothing to be afraid of. And when we go back to the anthill, you'll have a great adventure story for our little sisters. Oh, my name is Cira, by the way."

They found the trail left by the scouts exactly where they had said it would be, and from there the track continued along a very thin path through the grass. Lea was fascinated by lights dancing through the leaves high above the garden; the dapples of sunlight played with the shadows made by branches, a continual flicker to the beat of her steps, like something from an old silent movie. In fact, it was as if she was walking through a really worn old cartoon. She kept staring in amazement at the effect of the light as she moved along.

"The trail heads off this way!" said Ara, the leader of the group. "Follow me, girls."

They reached the edge of a clearing and came to a sudden stop, for from that point onwards the trail the scouts had left led across a long, yellowish plain. From their side, they could see nowhere to take cover if they walked across the field towards the stash of food.

"I don't like this one bit," muttered Ara. "Let's see, two of you go ahead and warn us if you see any danger. You, Cira and the new one... What's your name?"

"Lea."

"Okay, Lea and Cira, go on ahead and, if you see anything suspicious, warn us. But don't stop."

The hearts of ants are very small, but Lea's, which was still growing, was even smaller, beating a thousand times a minute as she crossed the chalky plain.

"Come on!" Cira encouraged, "move those legs, don't fall behind."

The instant she said this, a shadow began to fall and spread across the ground. It was shaped like black wings rapidly expanding above their heads in silence, leaving a cloud of dust in its slipstream that covered the silhouettes of the two ants. In the blink of an eye, a tumble, another tumble. Bing, bang, bong and then silence...

"Ow, ow, ow, what was that?" cried Lea, very much alarmed "Cira, Cira! Where are you? I can't see you."

When the shadow faded and her eyes adjusted to the light, she saw Cira lying motionless on her back, buried in a layer of dust a few metres away. Lea rushed towards her, with lots of questions."Cira, Cira, Cira! Can you hear me?"

After a few seconds that seemed like hours, Cira, still scared stiff from her tumble and trying to catch her breath, gasping as she spoke.

"That was the magpie. Saw it purely by chance right above my head. A little closer and it would've had me for a snack.

What an awful fright! By my antennae, goodness! Luckily, I managed to step out of the way at the very last sec. Let's get out of here!"

Eventually they reached the other side of the deserted plain, then hid under a clump of the tallest grass they could find, and from there sent out a message by making vibrations in the earth. Hitting forelegs against the ground, they made a rattling sound, informing their sisters about the situation they were in, and warning them of the danger in the air.

Once they were all together again, all of them safe and happy not to have suffered any casualties, they continued on their mission. They were now getting really close, with only a few steps left to climb to reach the smooth tiled surface of the porch at the entrance to the house. Even from the bottom step, they could hear Pavarotti's fine-feathered singing. Below the yellow prince's brass cage was a treasure trove of birdseeds, millet and hempseeds. When he wasn't singing, the canary stirred the feeder with his beak, tossing the less appealing seeds out through the golden bars, while he gulped the tastier, softer ones down his throat for breakfast.

"We made it, hurray!" Ara shouted as she raised her legs. "There's more food here than we can carry, so we'll have to tell the other gatherers to sweep up the rest later. Everyone take the biggest grain they can carry – no time to waste. Come on, follow me!"

Then, having spoken, she was the first to climb the steps. The others followed suit, imitating her like an Olympic acrobat team. Once they had reached the top, and everyone was holding tightly onto their seed, they set off back towards the anthill, but this time they decided to take a detour so that they would not have to cross the deserted plain where any enemy could see them.

This plan was not without its own risks, for in addition to the huge detour lugging the seeds on their backs, they would also have to cross the territory of the ants on the far side of the garden, those ants who had such an awful reputation and about whom terrible stories were told. Even so, they decided to run the risk.

3

The far side of the garden

The journey back became much harder; the gatherers had to walk with their loads across the desert on a hilly path that rose slowly through the gaps between the granite blocks that formed the wall. They zigzagged over a surface so rough that the movement shaved the hair from their bellies and legs. Exhausted, they stopped to rest on a bed of moss growing at the foot of the wall, in the shade of some weeds.

"Quiet! Someone is coming," warned Ara, who could feel a vibration in the few hairs that remained on her legs; at first, this was no more than a distant sound, like a crowd gradually drawing closer. After a few minutes, it was beyond question: this was the steady march of an army of legs approaching, causing the earth to tremble in a way that made their exoskeletons rattle.

"Hide!" ordered Ara.

The ants scattered, squeezing into the very first nook they could find. Some of them climbed up on peony buds that were about to bloom. So quietly, they waited in their hiding places until they saw the anthill troop appear from the far side of the garden. They arrived altogether, in such a long, long line, marching at a brisk pace, chanting army marching songs, boasting about the foul deeds they had done throughout their history.

"Aren't they full of themselves? As if these were exploits to be proud of," thought Lea from her hiding place, not taking her eyes off the parade. As they passed, one of the last ants suddenly turned towards a sandhill – something had caught its eye.

Ara had hidden in the first crack she had seen, in a small mound of earth that the wind had piled up against the wall over the years. But the hole was so shallow, when she backed into it, her head and front legs were left exposed. In her attempts to hide from view, she pushed herself down into the hole, wiggling her hindquarters about to make room. But with all the wiggling, she accidentally kicked over some pebbles with her front legs, and they rolled down in slow motion, causing a small avalanche that ended with a crack.

"Alert!" cried the ant at the rear. "Alert! Alert!" she crazily repeated again and again.

"There's an enemy hiding up there!"

"Be very careful! Keep your antennae up, maybe she's not alone!" warned another ant, who seemed to be the leader of the group.

A group of ten or twenty ants broke away from the formation, which carried on marching as if nothing had happened, and immediately the warriors were standing in front of the hollow where Ara was crouching.

"Come out, stranger!" commanded the chief, who was the tallest of them all. "Don't make us drag you out in pieces."

Left with no way out, and so as not to betray her sisters, Ara climbed out of the hole with her front legs up, as dirty as a dung beetle in a stable.

"Here I am," she said calmly. "I have done nothing wrong in these lands. I took a wrong turn and I'm lost. Could you help me?"

"Then why are you hiding?" the chief replied. "You were watching us. I've no doubt you're a spy, and the punishment for spies is death. Seize her!"

Immediately, several warriors leapt at Ara like demons, grabbing her legs with their claws, and when they had her pinned down, another, uglier than the shadow of a dead ant, bit into her antennae and pulled hard to tear them off. Ara's sisters, not being able to bear the scene from their hiding places, could take no more of it and suddenly leapt out of their hiding places, their bodies puffed up to frighten off her attackers. If they wanted to free their leader, they would have to fight for their lives as well. But their enemies were not caught unawares: another platoon was lurking in the background, ready to counterattack if there were more invaders in hiding. In an instant, a fight to the death broke out.Lea had barely stepped out from her hiding place when a much taller soldier with a face like a chickpea grabbed her by one of her antennae with her huge jaws. Seeing herself trapped, Lea leapt back in reflex, and driven by the sheer force of fear and fright, took such a jump back that she was catapulted into the air like a wire doll, and as she flew, she kept jumping; she seemed possessed by St Vitus' dance, the way she jumped up and down, frightened by everything, as if being chased by her own shadow. This dance left the warrior so baffled that she froze, with her hindquarters stuck in the sand, staring off into the distance while rubbing her antennae, not understanding what had just happened. Even Lea didn't know how she had done it, but fortunately, it had worked: she had managed to escape by the skin of her teeth, which were chattering with fear.

As she scurried through the undergrowth, with dirt in her eyes and her antennae drooping, she caught a glimpse of the barbarians finishing off her sisters.

Ara, Cira and all the others were lost! There was no way left to help them. She hid behind a smooth stone over the wall and from there watched the ants move across the garden, carrying their precious canary seeds and leaving behind the lifeless bodies of her older sisters. Scared stiff, she wept and mourned their loss until she fell asleep in a crease in the bark of the blue olive tree.

As the sun was coming up, Lea plucked up the courage to stick her head out of her hiding place. Sensing no threat and hearing nothing suspicious, she crept out of the blue nest where she had spent the night and with great stealth made her way to where the bodies of her sisters lay. They seemed to be asleep... Carefully, so as not to wake them, she placed them one by one on a bed of wild laurel flowers that had fallen next to a rounded corner.

When she had finished saying goodbye, she realised that she didn't know where she was or how to return to the anthill; she had no idea which way to go. She looked in every direction, waving her antennae about, yet she couldn't make up her mind. Finally, she began to walk aimlessly, following the whistle of a breeze sweeping over a field of newborn daisies.

4

Alfi

Lea wandered aimlessly for days without knowing where she was going. She wasn't able to eat anything and still felt very frightened. She hadn't come across any other bugs either, until that afternoon, when she spotted a long, motionless, pebble-like figure in the distance, resting quietly on the stem of a gladiolus. It was a very strange creature, and looked really quite big. Little Lea had never seen anything like it before. Sitting very calmly on its two long hind legs, it seemed to be waiting for nothing. Its antennae sprouted eagerly like the branches of a dry tree above its eyes, which were a pair of oval globes that resembled welder's goggles over the sides of its head. There was a vegetative stare to the eyes, the same look that toad-skinned melons have when they see us passing on the road. Its shell was hard and rough on top, with rows of hooked spikes ending in sharp points. The rest of the body looked smooth, like a polished stone, lined with seams the colour of stagnant water.

"Hello," Lea greeted after watching it for a while.

When there was no reply, she addressed the strange bug again from a safe distance in the deepest tone of voice she could muster.

"Who are you, and should I be afraid of you? If I come near you, will you attack me?"

After a long silence, the creature began to speak in a slow voice that sounded like it was being blown through a bagpipe, repeating a few words as if the echo would speak for it too.

"My name is Alfi, Alfi." And, after a long pause, she continued: "That's a lot of ant questions for one answer, answer. You have nothing to fear from me, me. I am a vegetarian, I only eat grass and grass shoots. Not that you can trust, trust, locusts in general, but you can trust me in particular, yes, yes. I'm pretty peaceful, and I don't normally, normally attack ants. I fly, fly as the wind blows, from a place where the earth, the earth ends, launching myself into the air with my hind legs, hind legs, spreading my wings, transparent wings, that keep me hanging in the air, the air, these are my parachutes, rising on the breeze, breeze. The strong wind over the last few days has finally separated me from my nagging companions, companions, and here I have landed, in this garden with no name, no name."

After another pause, she asked her:

"And you, what is your name, your name? Are you travelling alone, alone? Where is your family, your family?"

"My name is Lea, I am alone and lost. Lost, lost," Lea replied, adding the echo to make herself look good.

"Lea, Lea," said Alfi in her breathy voice, "you don't have to repeat the words like me, me. I understand you perfectly well, ant, ant, from the way you talk."

“As you wish, I didn’t mean to offend you, I just wanted to be kind and answer you as you speak,” said the ant. “I don’t have much to tell you. I don’t know how many days I’ve gone without seeing anyone. I lost my sisters in a fight when we were attacked by ants from the far side of the garden, and I don’t know how to get back to my nest. I am hungry and have no idea where to go.”

After a long silence, the locust spoke again:

“If you don’t know where to go, any way will do, will do. You can stay here with me if you want, if you want. This is a fine place, so you’ll be safe, safe. There’s enough grass here for us both, us both.”

Lea stared in amazement at the row of little hairs hanging from the bug’s chin, falling down like the fringe of a quilt. It was as if the row of hair had a life of its own. The little hairs danced back and forth together to the beat of her voice, making the wave and laughing their heads off every time the locust echoed her words. Lea found them so funny that she decided to trust Alfi, for the time being at least.

"Thank you, Alfi, I'll stay with you to get my strength back."

For several days, Alfi told Lea many things about her life, such as why locusts did not live in a house like ants. Once they were no longer nymphs and became adults, they flew in swarms of millions, like a sandstorm so big that when they flew in front of the sun, day turned to night as during an eclipse. The day the green-black cloud flew across a crop field, all of them landed at the same time and stayed there, hopping about; they swept up anything that was edible, and when they left, they left a lifeless, dusty desert. For this reason, the locust plagues were no laughing matter for anyone, except for the birds that chased them, which grew very fat by hunting down mountains of them without effort, as they flew about confidently; but as there were so many of them, the losses were scarcely noticeable. She also confessed that they were not as friendly with each other as the ants were, to those of the same anthill, of course. Locusts interacted as little as possible with each other, just enough to get by, but they were really quite unfriendly.

"That's why I'm so calm now, calm, without having to share the air with millions of selfish, selfish vermin," said Alfi, expressionless except for the little hairs on her chin that were dancing at their will.

Lea lowered her head and said with sadness in her antennae:

"I'm not. I miss my sisters too much. We ants can't live long without each other's company because we are all part of the same body that lives and works for the anthill. This is how we are."

"That's fine, fine, but I wouldn't fancy, I wouldn't fancy living as an ant, always being so dependent on all the others, the others. I prefer to be free, free. Besides, not all ants are like that, like that," the locust shook her head.

And she began to tell her the story of a young queen ant she had met on one of her solo landings.

5

The Abandoned

Her name was Ora, and she was the largest of all the queen ants born that year. More than twice the size of her sisters, she had four silver wings running up and down her back. Ora told Alfi that the mother was well known in the surrounding fields, especially for her cruelty to her worker daughters. She had a court of warrior ants who protected her and were also the guards who kept order in the anthill. This queen was feared because any creature that strayed into her domain was hunted down and imprisoned. When they had gathered many captives, they would devour them, dead or alive, at great feasts organised by the warriors in honour of their mother queen.

The queen laid the eggs of each generation. When the offspring were old enough, they were forced to slog relentlessly through the hot months until the anthill's foodstores were fully stocked, and when the first frosts bit, they were left to starve. Then, when the next generation was born, it was back to square one.

The guards didn't mess around. They forced the workers to work from sunrise to sunset; if they stopped, they were punished and forced to carry more weight than they could bear. Hundreds of innocent ants died to set an example for the others.

When Ora grew up, she was ordered to leave the anthill, to fly away and start her own family. But she wasn't interested in becoming queen, or having her own anthill, or being far away from her sisters. The guards ran her out of the anthill several times. The last time, with the queen's permission, they threatened to execute her on the spot if she ever showed her face again. So, she had no choice but to leave for good.

One morning, as usual, the worker ants were forming their very long line, one line quickly filing out of the anthill and another returning in procession with their heavy loads in tow. They saw a huge queen ant approaching; she looked like a black olive, shining like an obsidian mirror. She was accompanied by a retinue of strange creatures, a troupe of stray critters parading haphazardly in the sun like drunken gypsies. Together, they created a spectacle unlike anything the anthill workers had ever seen.

A crazy grey moth led the way, circling above the heads of the other creatures in her gang, occasionally perching on the trunks of pine trees to wait for the others to catch up, then resuming her mad flight to the laughter of the group. Following were a host of ants from different families: some red, some white, some grey, some black – ants of all breeds and sizes. They were joined by crickets, earwigs and many other bugs in a jumble of insects so chaotic it was both ridiculous and hilarious. They walked and stumbled around the Olive Queen; that is, they walked as best they could, for many of them were missing legs, or had their antennae cut off, or had misshapen bodies. They looked like a band of maimed zombies with a purpose.

A group of albino lice, rather overgrown for their age, pranced through the air without crashing into one another. They danced with a steady tropical frenzy, mocking a shaggy bumblebee as they passed in mid-air. Monda was at the tail-end of the parade, a large, blind, six-eyed spider who walked heavily and carried on her back a group of six very lively fleas that had escaped from a circus and now worked as the spider's guides. They were very clever and spoke every language.

All these creatures formed a strange family which they called 'The Abandoned'. In some ant hills, you would hear talk of them. In every public place they came through, they were cast as a lazy, worthless bunch – a bad example for any hard-working, orderly society. This was because they obeyed no one and lived free from social ties, without a land or obligations. After all, most of them had been cast out of their anthills as useless, or were escaped convicts, or had been lost and half-dead when they were rescued. So they were used to rejection, and wherever they went, their way of life was seen as a threat by leaders and most middle-class ants. It is also true that they were secretly admired by many, and so the queens tried at all costs to prevent their obedient subjects from finding out there were other ways of living and thinking outside the discipline of the anthill. But no one imagined that being free meant a very difficult life, a life that required many sacrifices. In fact, in all the anthills they passed through, it was the most submissive, miserable ants who picked on them the most and showed the greatest contempt towards them.

When the bizarre entourage encountered the line of worker ants, the guards reacted immediately and approached the group. The leader stepped forward and said:

"What are you clowns doing here? Go away before we lose our patience and make soup out of your shells." The Olive Queen rose up on her hind legs, displaying her entire body, and puffed out her chest. "I wish to speak to your queen; you are not of a high enough rank for me, a queen, to address you."

Her voice echoed through every road inside the anthill.

"Listen to what at she's coming out with now! The queen? Who do you think you are to speak to her?" said the head of the guards with contempt.

There was a deathly silence. The ants in the line stopped what they were doing and were staring dumbfounded at the scene. Then the Olive Queen spoke again, addressing them all at once.

"I am a queen ant and I look after all my fellow ants, regardless of where they come from, what breed or species they happen to be. Together, we form a family and have no obligations. You, too, are free to live as you wish. Listen to me, all of you: those of you who wish to join us are welcome. Now, call your queen," she said, looking at the guards, and added, defying the chief with her antennae, "I will not say this again."

The guards were used to spreading terror and being obeyed by the ants without a word of complaint, so the Olive Queen's command came as a surprise. They exchanged worried looks. Suddenly, out of the blue, four of them leapt at the Olive Queen with their jaws wide open, grabbing her by the antennae and her legs, ready to teach her a lesson in front of everyone. But the queen, undeterred, quickly took a step back as though it were part of a dance, and in one swift movement shook them off with the back of her abdomen, like a dog shaking off fleas. The guards were thrown a great distance, dropping in all directions one by one, as if a gust of wind had unleashed a shower of broken branches.

"Alert! Alert! Alert! We're under attack! All hands to battle stations!" shouted the head guard.

Chaos broke out at the entrance to the anthill. Those rushing in crashed into those rushing out, who were knocked down like bowling pins. The guards were so shaken that they were swiping their way through the workers waiting to enter, and ended up scattering their loads all around.

"Silence!" came a voice from underground. They all froze like mosquitoes trapped in an amber rock. "Guards, to my side! Whatever is all-this fuss about?"

The voice came from inside the anthill and through the gaping hole in the entrance appeared jagged jaws like bread knives. Like a train emerging from a tunnel, a beetle with a shiny black bean-shaped head appeared, followed by an armoured body.

The armour was made up of separate moving parts attached to one another, all stitched together in the middle. Outside the anthill she moved gracefully on her six, gigantic hairy legs, her armour dazzling everyone. She was the Mother Queen.

Furious like a samurai with chronic gastritis, she approached the Olive Queen's group and, surrounded by all her guards, raised her voice:

"How dare you disturb the peace of my anthill, don't you know what we do with foreigners?"

"I am not a foreigner," the Olive Queen replied. "I am Ora, your daughter. And we do know what you do to those who fail to obey your orders."

The silence was broken by the startled whispers that began to spread among the worker ants.

"I haven't come to look for trouble, I was just passing by with my sisters. I stopped to greet the family and to offer a place of freedom to those ants who wished to join us."

"My queen-born daughters can never return; there is only room for one queen here. We told you that if you came back it would end badly for you. But not only have you come back, you have come back to stir up my anthill with your ridiculous ideas. No one here wants to go with you."

And looking at her worker ants, she said:

"If any of my daughters are keen to join this bunch of useless outcasts, they may go, I won't object."

But none of them moved from their places.

"You see, Ora? No one wants to go with you to live like an ant without a job or standing. None of my ants can survive on their own outside the rule and protection of their queen."

"Listen now to me," said the Olive Queen. "Do you know what will happen when the frost comes? Do you think the food you bring in each day and leave in the stores will be enough to get you all through the short, cold days? When winter comes, there will be no food for you at all; it will all go to the queen and her court of murderers, who will save the bare minimum to feed the larvae of the next generation until they are ready to work when you have all starved to death."

None of them said anything.

"Very well, you asked for it," said the Mother Queen. "Attention, my guards, form an attack position, we are going to wipe out all these intruders."

6

The battle

All of the guards attacked, and a one-sided, ruthless battle ensued.

'The Abandoned' were not strong enough to fight them off. Those who could not flee or hide fell at the first attack. The lice jumped on the heads of the larger ones, twisting their antennae, and immediately jumped back to take cover until the next attack. Monda the spider, guided by her fleas, shot silk threads at the attackers' legs. When the sticky thread was nicely tangled, the moth would pass the thread over the twig of a bush and Monda would hoist it up to dangle in front of the birds. The Olive Queen flung any warriors who rushed at her into the air, leaving a trail of ants behind her until her antennae clashed with those of the Mother Queen.

"Madam, this is not what I wanted, but you give me no choice. Prepare to become a shadow."

Gripped by their jaws, the two queens tested each other's strength and began to roll about on the ground, slowly at first, slowed by the tension of the forward thrust. Then, little by little, they turned so violently that they created a whirlpool, swirling like water in a funnel, and finally became an uncontrollable centrifugal force of shell and legs, sweeping everything in their way. The young queen had gained advantage. With her last drop of energy, she managed to stop the inertia of the whirlwind and stand over her mother. She was about to deliver the final bite when the frightened Mother Queen, seeing herself defeated, kicked out desperately with all four hind legs, trying to shake Ora off. She kicked so hard that she lost a leg in the struggle and dented her breastplate. With one last effort, the mother rubbed her daughter's antennae in the dirt and managed to free herself from the youngster's grip. As she let go, she took to her feet – well, all but one – and staggered off to hide somewhere deep inside the anthill.

Some worker ants helped 'The Abandoned' and others continued to fight them, but when they saw their queen flee, they were astonished; they had never thought that their leader would leave the battlefield in such a humiliating manner. And so the warriors, seeing the Mother Queen flee in terror, gave up and followed her underground. It was all over. Victory was the Olive Queen's.

"The Mother Queen is gone, you are free now," said Ora.

But the anthill workers had no wish to leave their home, so they asked the queen of The Abandoned for a favour: to spare their lives and leave the place, as they had brought them enough trouble.

Despite all the misery they suffered in their anthill, they didn't want to leave; they knew what awaited them, but they were confident that one day everything would change. They hoped that some future generation in the anthill would have a just queen to look after them.

"Until that day came," said Alfi at the end of the story, "the terrible punishments that the daughters suffered, suffered at the hands of the Mother Queen and her minions, did not matter, the workers preferred to maintain order and keep up the routine, routine that their tyrant ordered them, rather than freedom, freedom which means being masters of their lives, their lives. They were afraid to improvise and to face the unknown, the unknown every day. That's why I can't understand the ants, the ants."

Thinking about the story of Ora, Lea wasn't sure if she understood all the parts, but she knew she couldn't help being what she was: an ordinary ant who felt very lonely without her sisters. So, a few days later, she said goodbye to Alfi the locust, truly grateful for her help and everything she had taught her. She left feeling more confident, even a little wiser, than when she had first arrived. And without looking back, she set off to find her sisters.

* * *

No one knows if Lea ever made it back home.

Or perhaps she crossed paths with The Abandoned and joined them in their adventurous life. Whatever did happen to Lea, what we do know is that her story has reached us thanks to the ants' ability to communicate. Thousands of stories are told every night anthills around the world, stories and tales passed from insect to insect. Each one telling it in their own way, and so the stories continue to change, like the game of the broken telephone. Just as Lea's story was told to me by someone who listened to the song of crickets, there are those who, if they listen carefully, can hear another story – or perhaps even the continuation of this one – in the calligraphy of fireflies in flight.

the end

Text editing
Antonia Castaño
César Álvarez

English translation
Jesús M. Silvestre

English copy-editing
Stephen Beechinor

Design
Sonia Sánchez and Fernando Maquieira

Cover Design
Maricris Herrera
Taro Cantú
@Estudio Herrera

Ant illustration
Milimbo

Preprint
Maq studio

Printing and binding
Liberdúplex

Tipography
Certau (APFEL Type Foundry)
Signifier (Klim Type Foundry)

Paper
Geltex
Arena ivory Smooth
Oria blanco

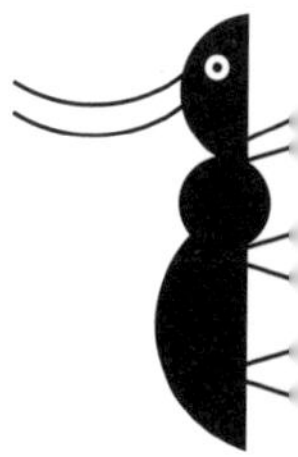

Editor-in-chief
Ramón Reverté

Editorial Coordination
Julieta Averbuj
Patricia Reverté

RM Verlag, S. L.
Loreto 13-15, Local B
08029 Barcelona
España

Editorial RM, S. A. de C. V.
Ciudad de México
México

info@editorialrm.com
www.editorialrm.com
@editorial_rm

RM #483

ISBN 978-84-19233-72-1
Depósito legal B 17301-2023

Printed in Spain 2024

You were always asking
me about the ant book.
For you, Dad, wherever you are.

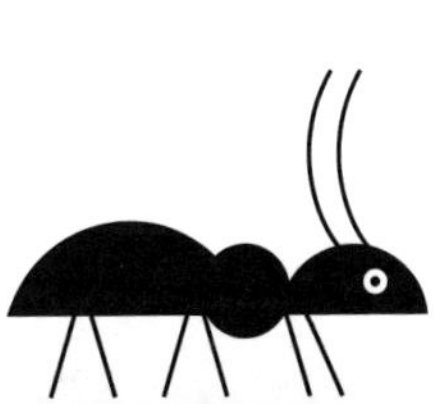

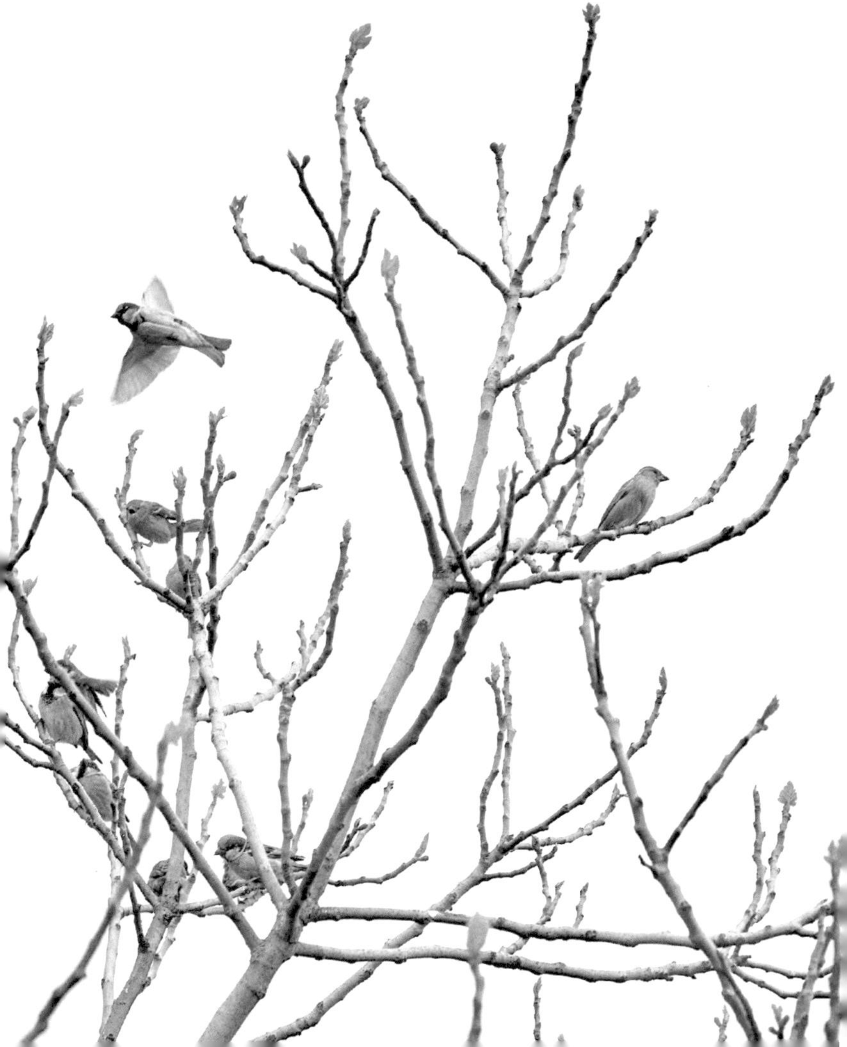

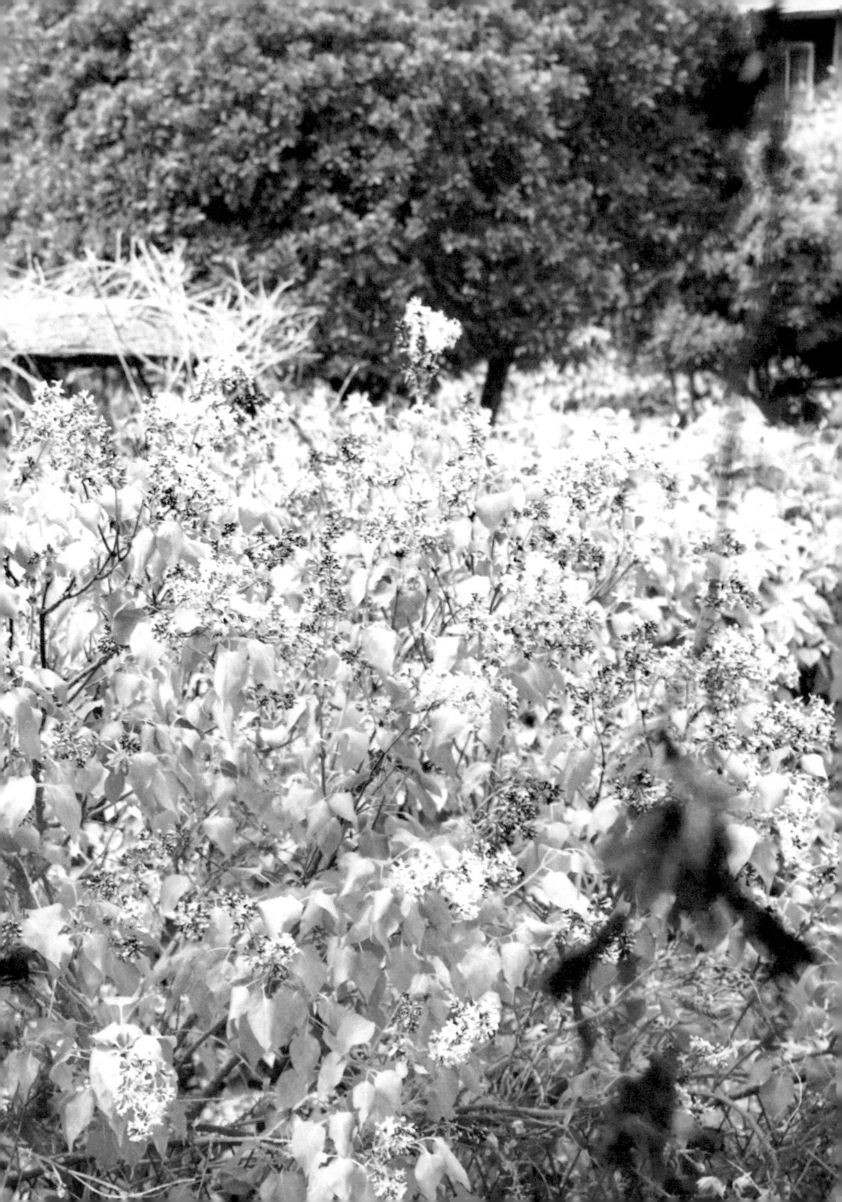

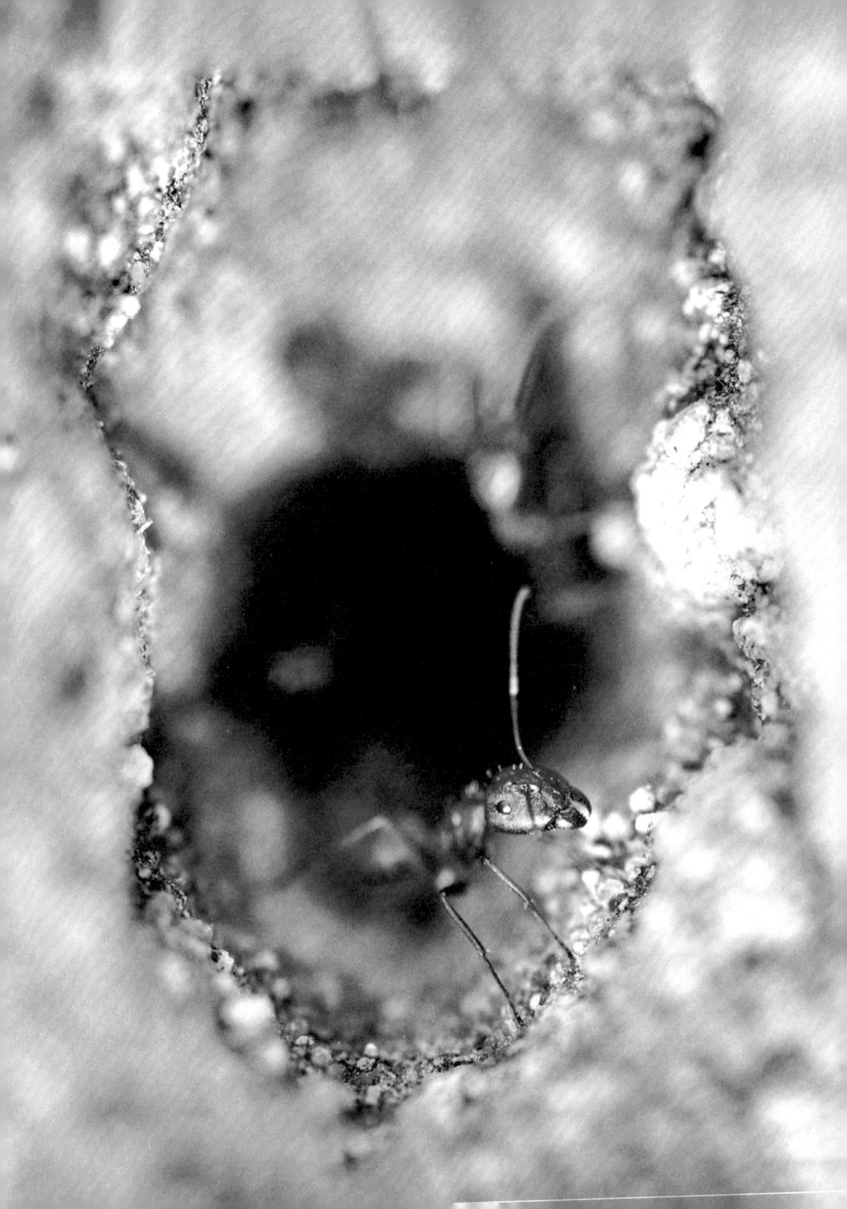

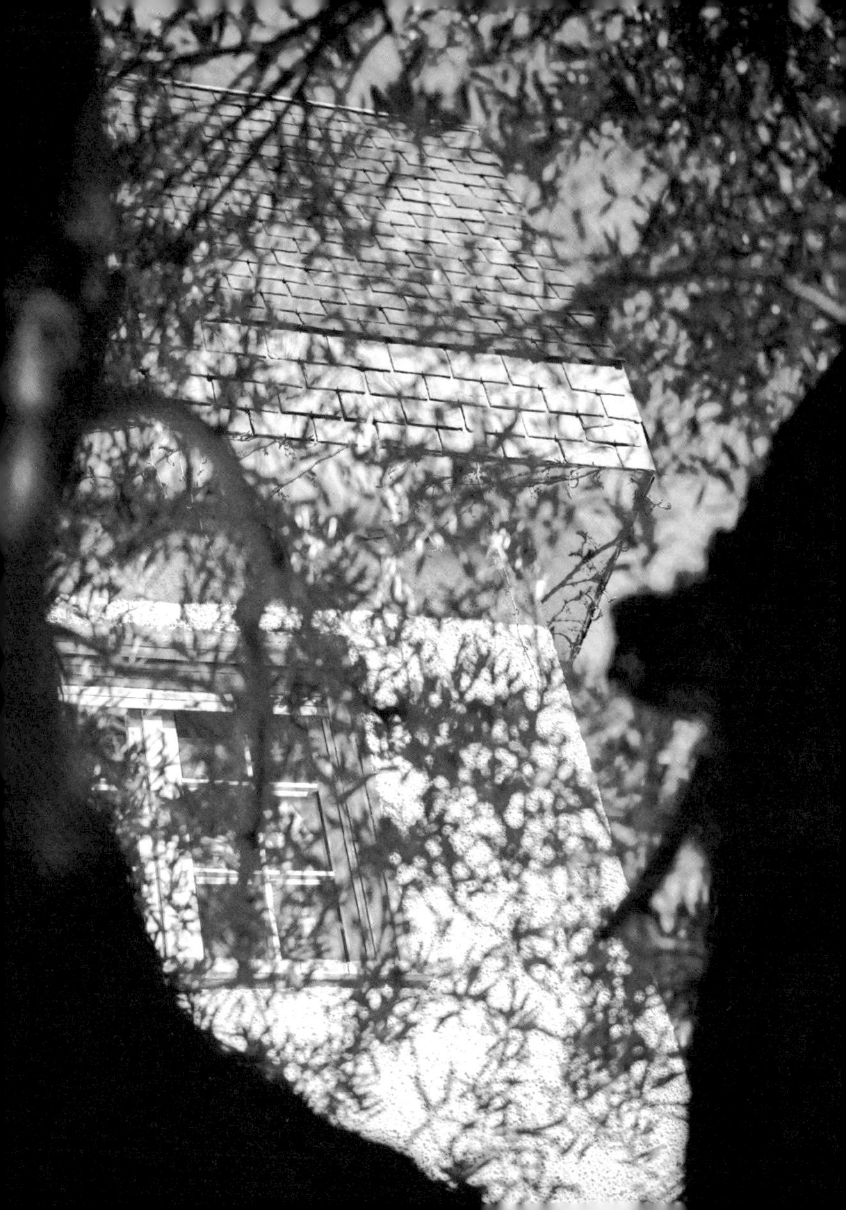

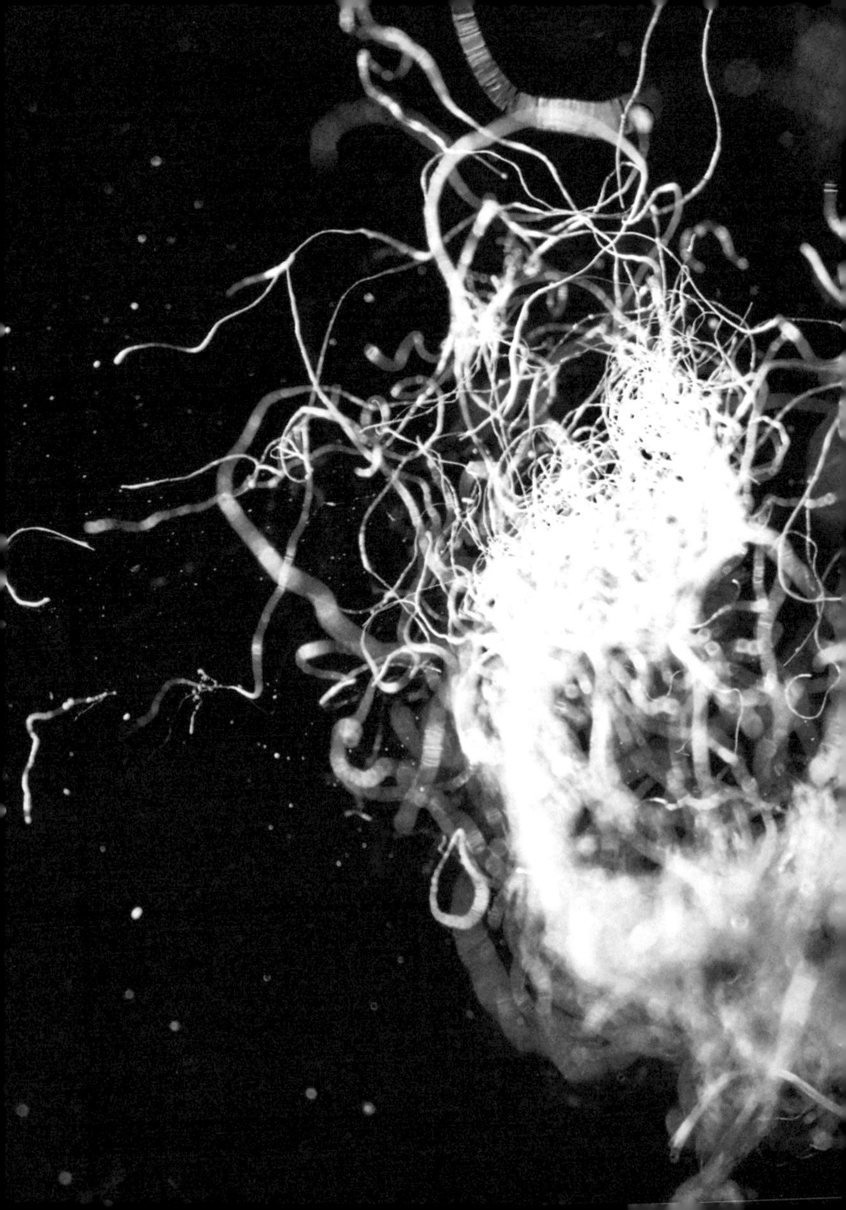

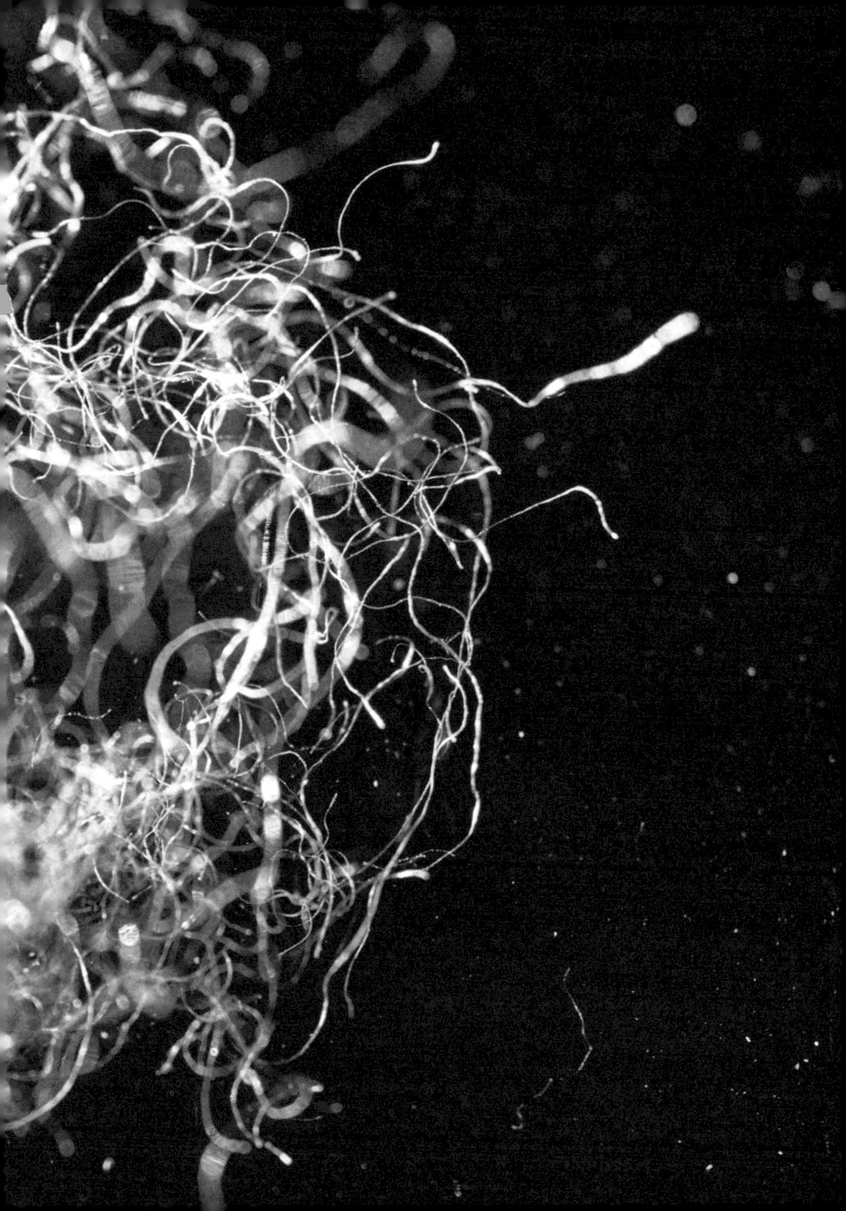

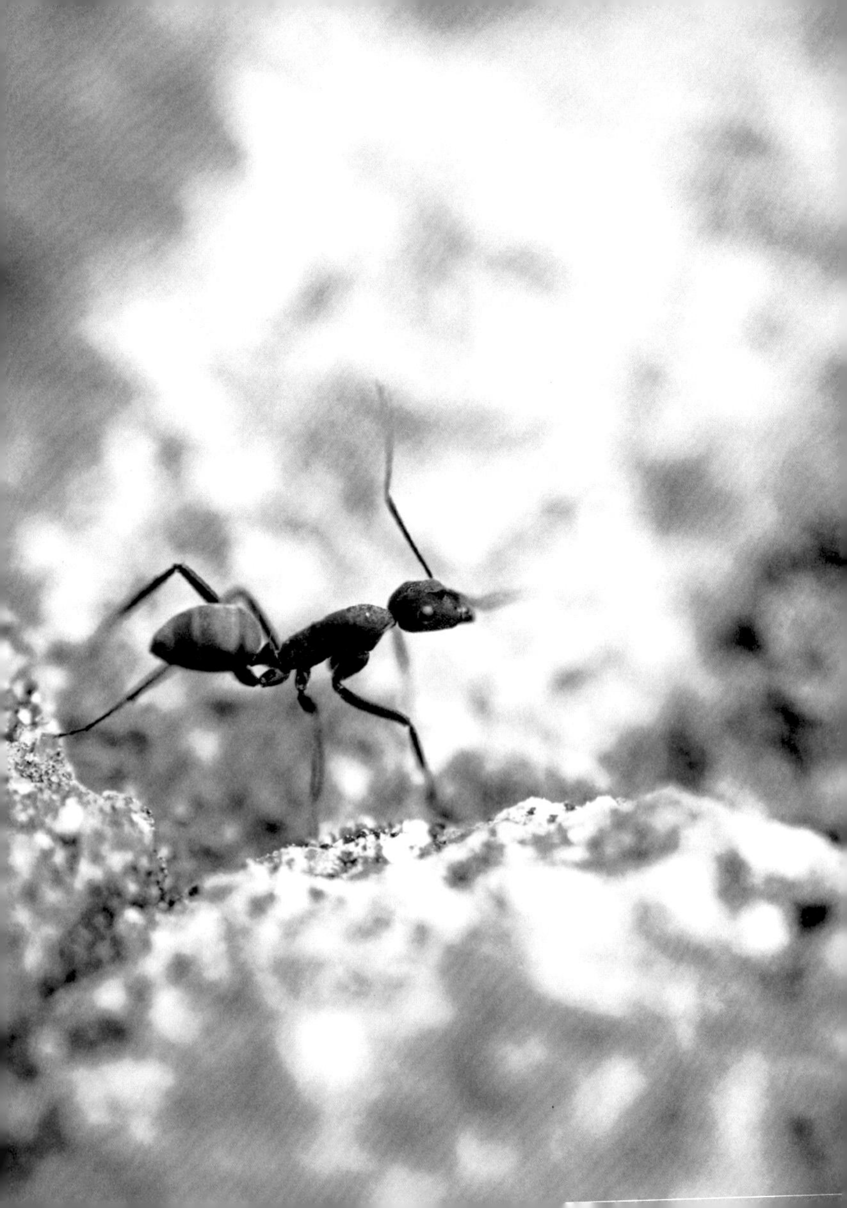

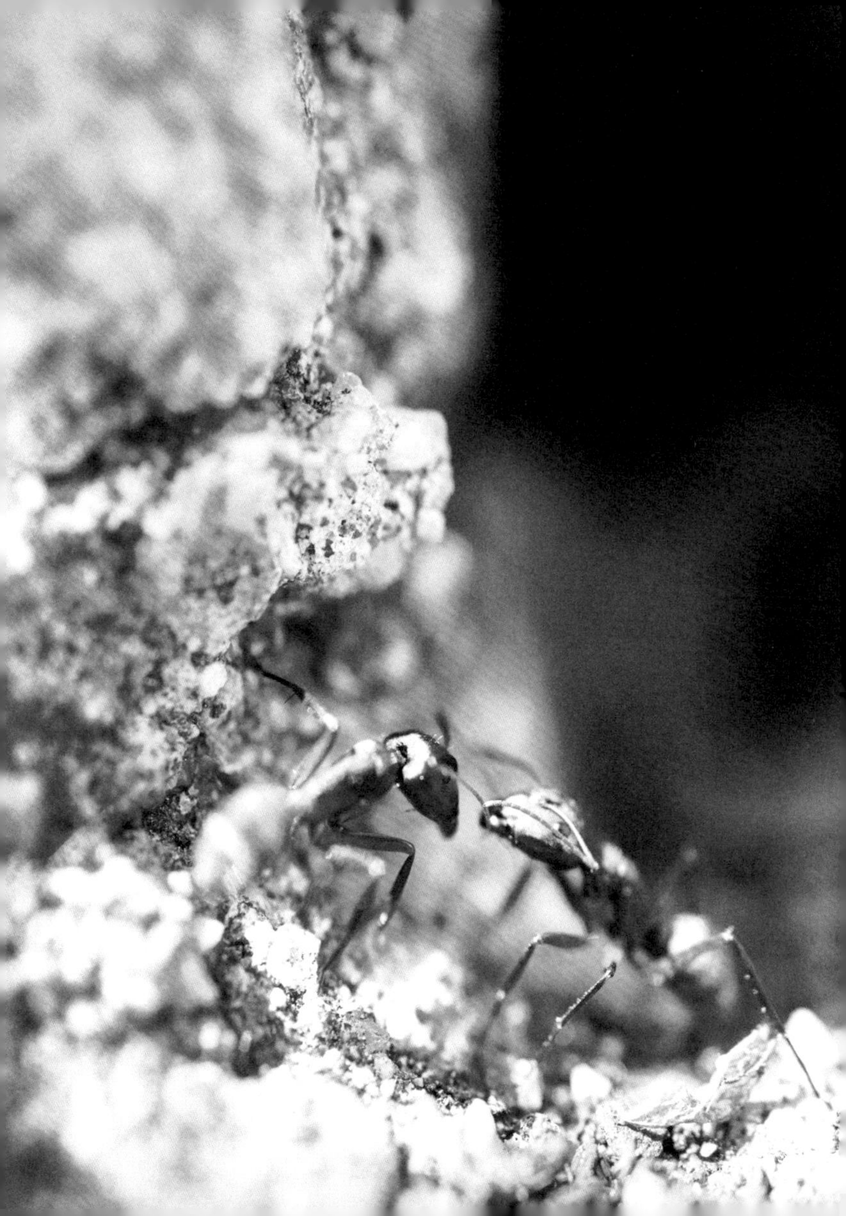

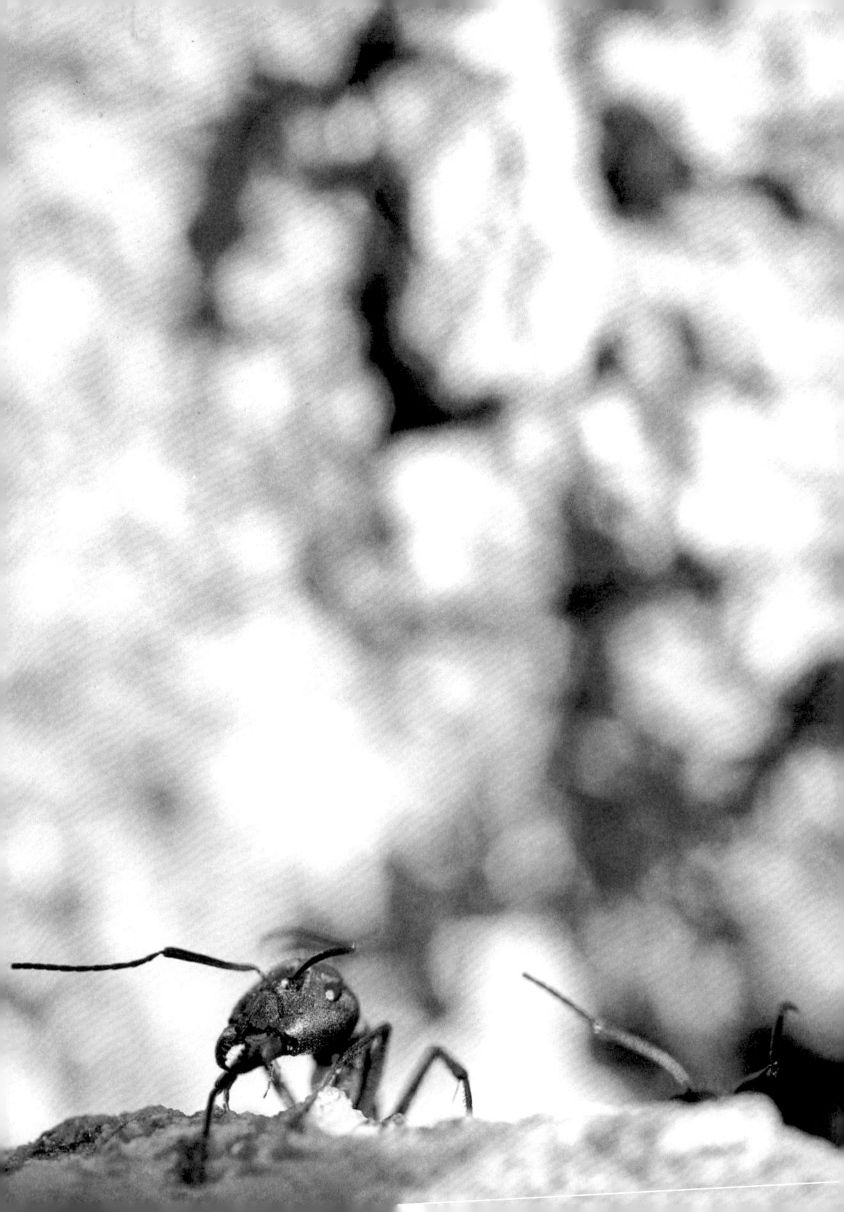

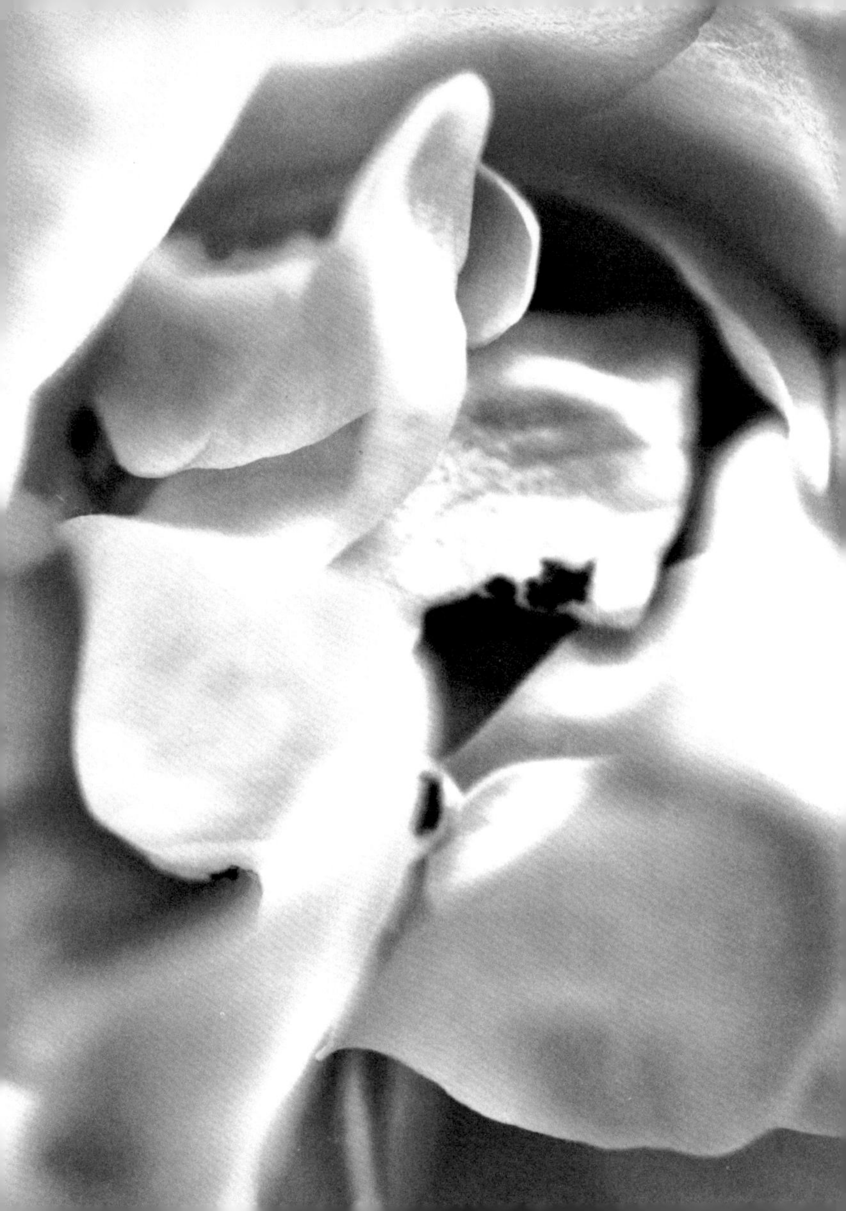

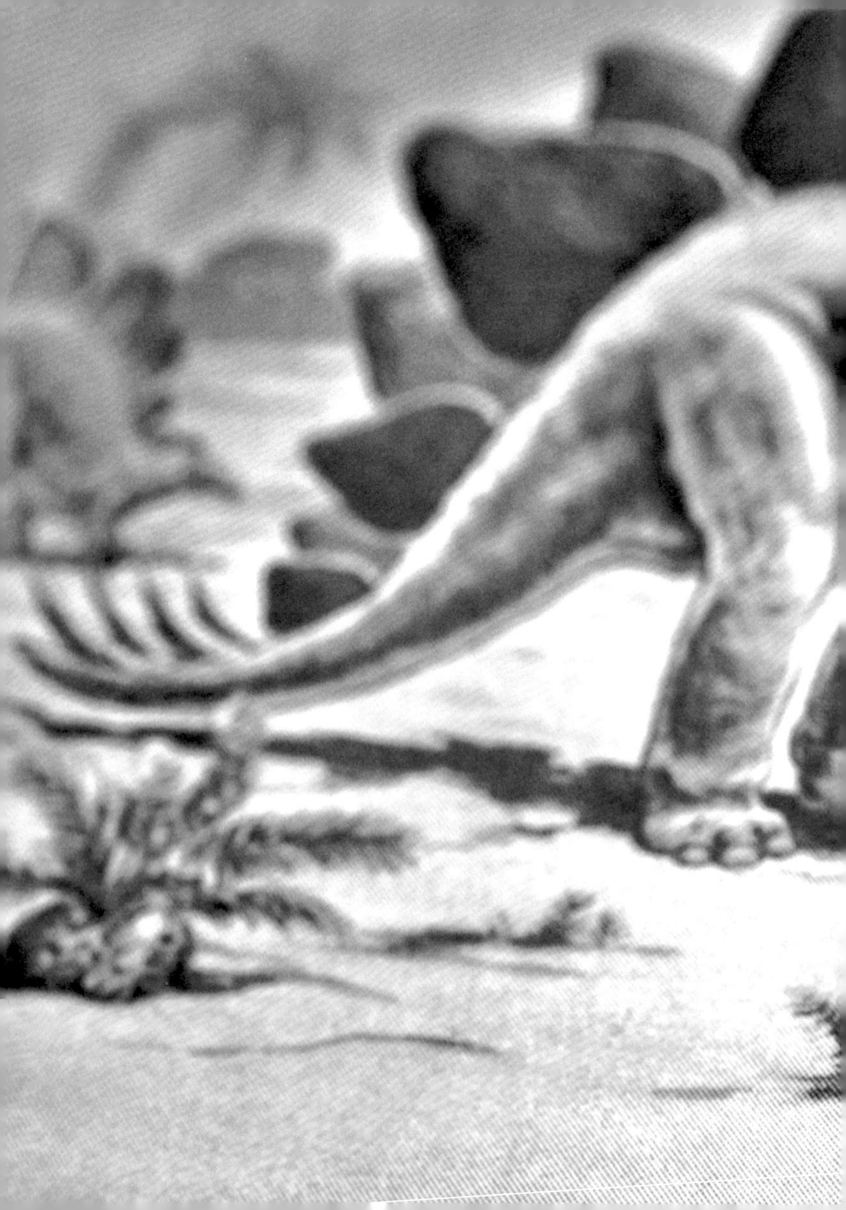

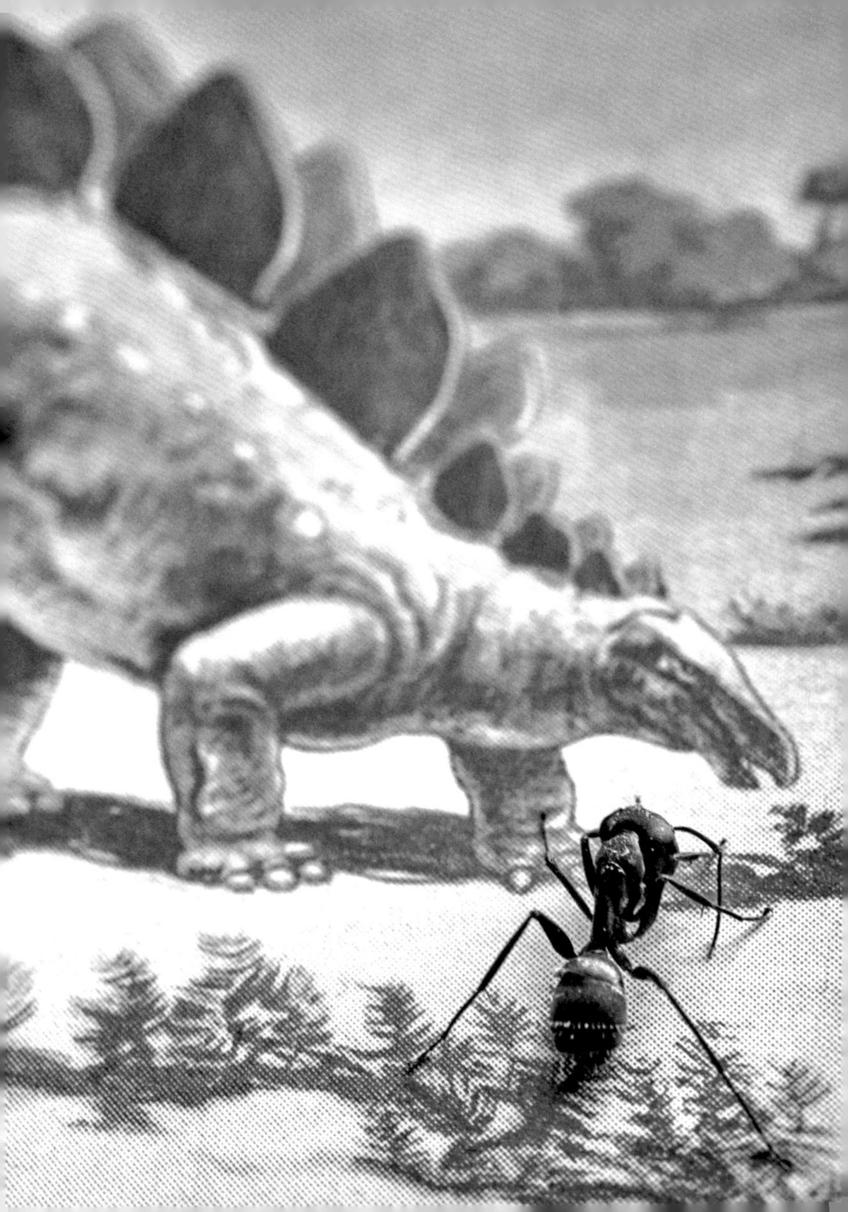

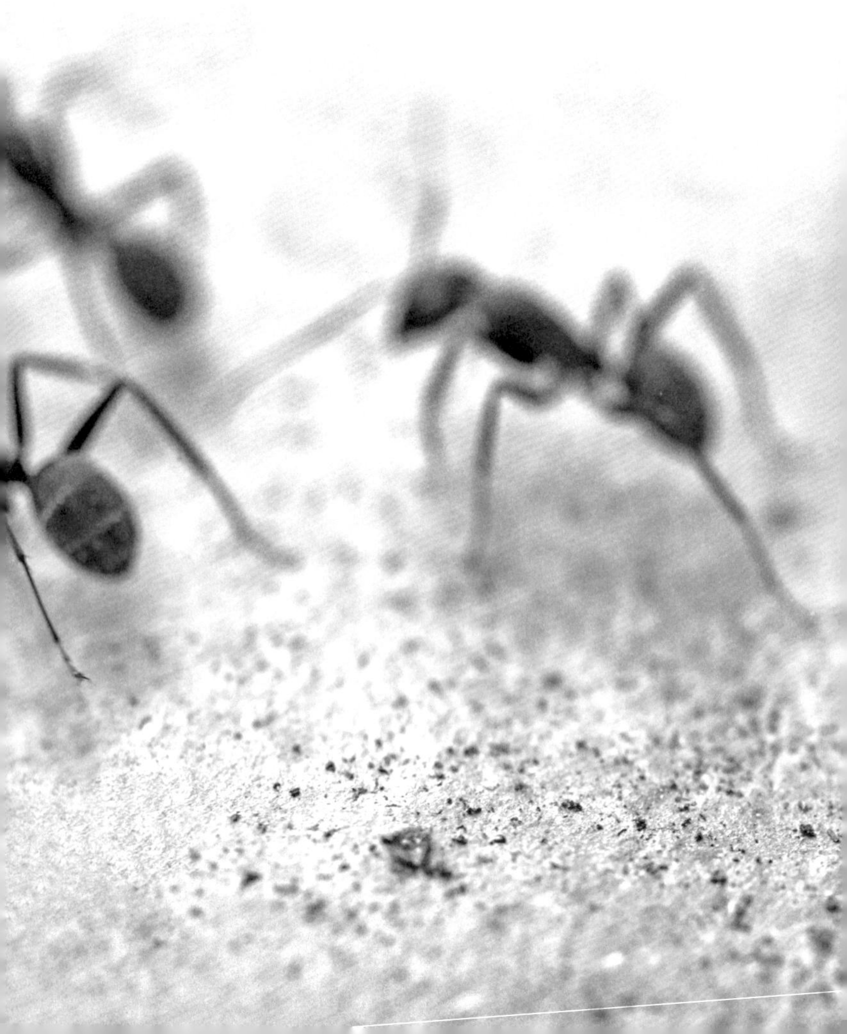

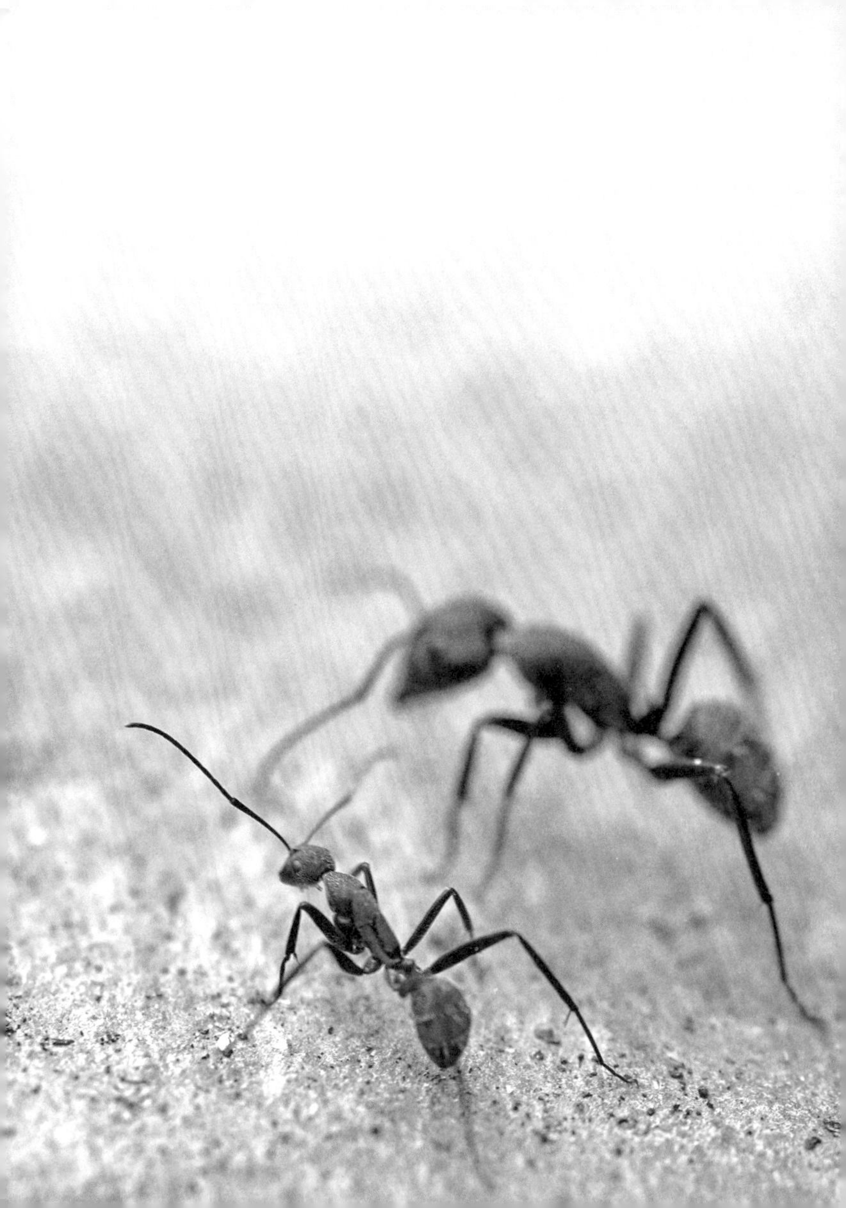

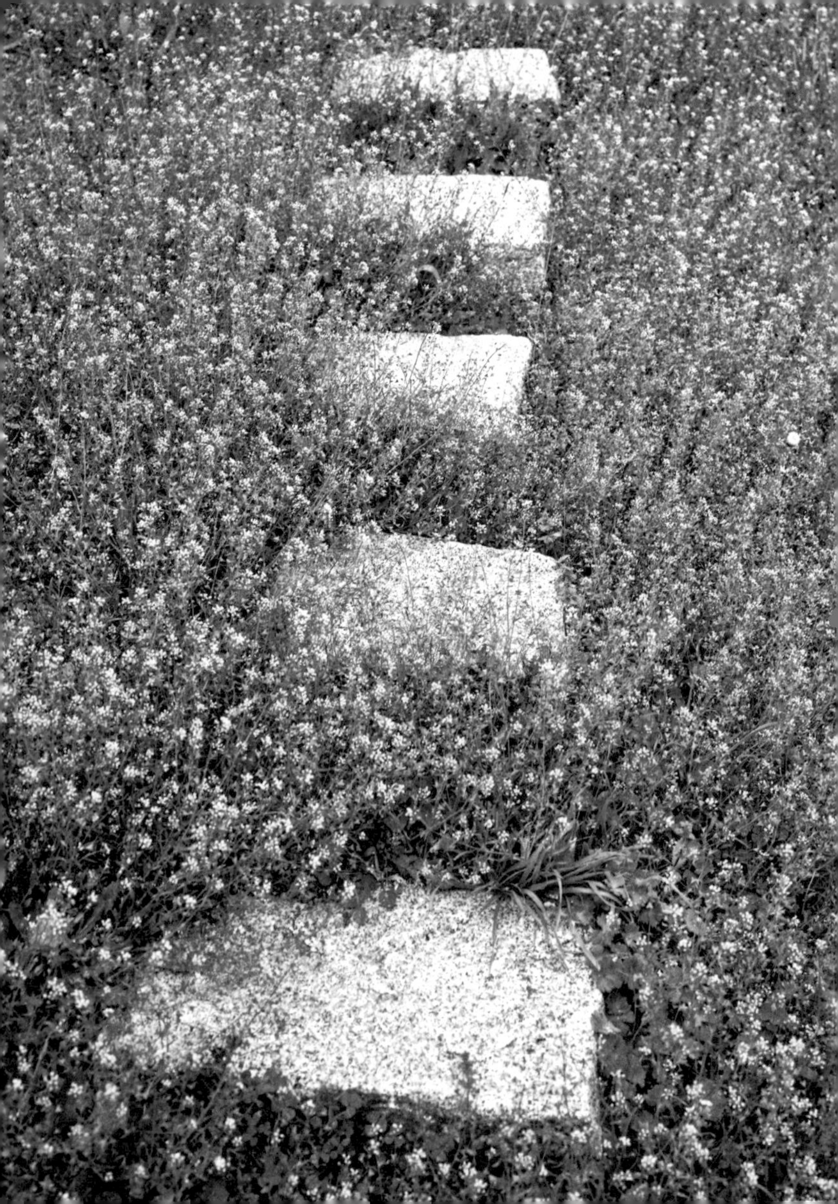

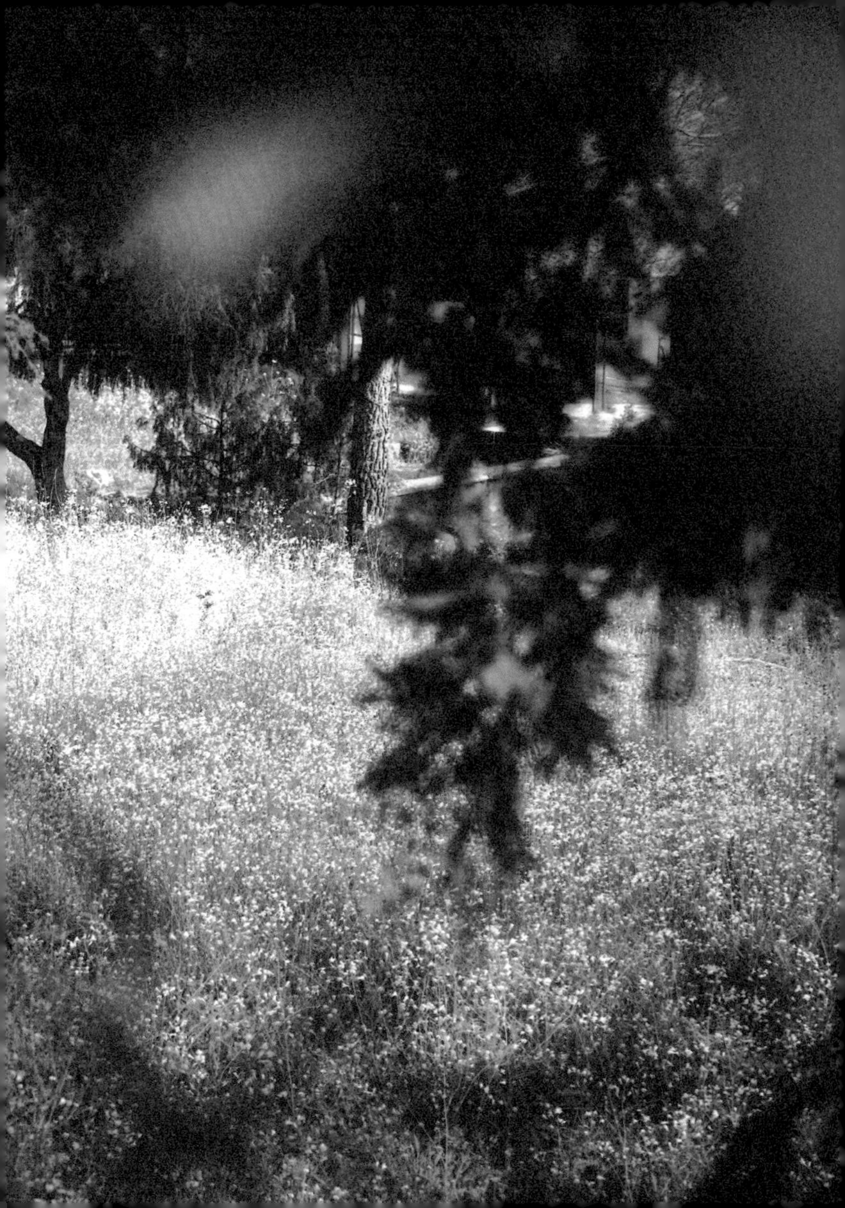

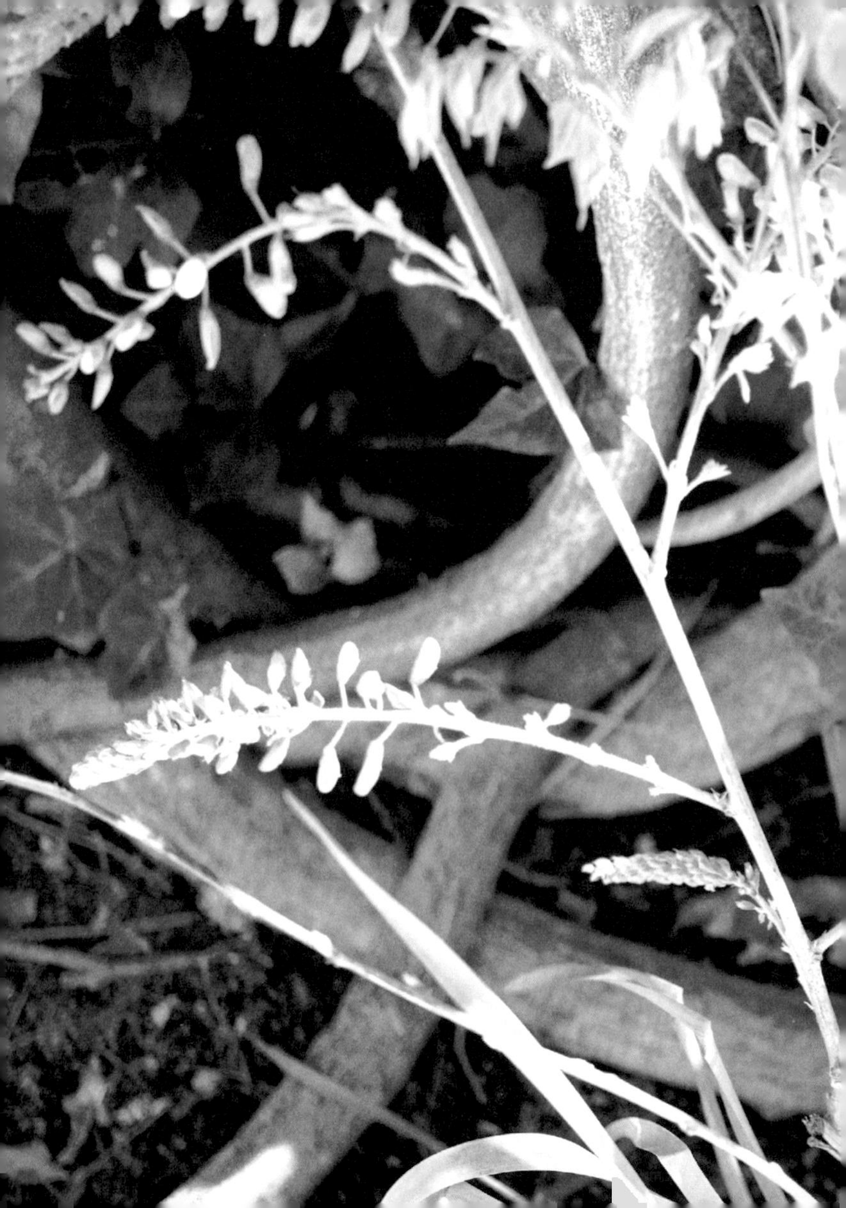

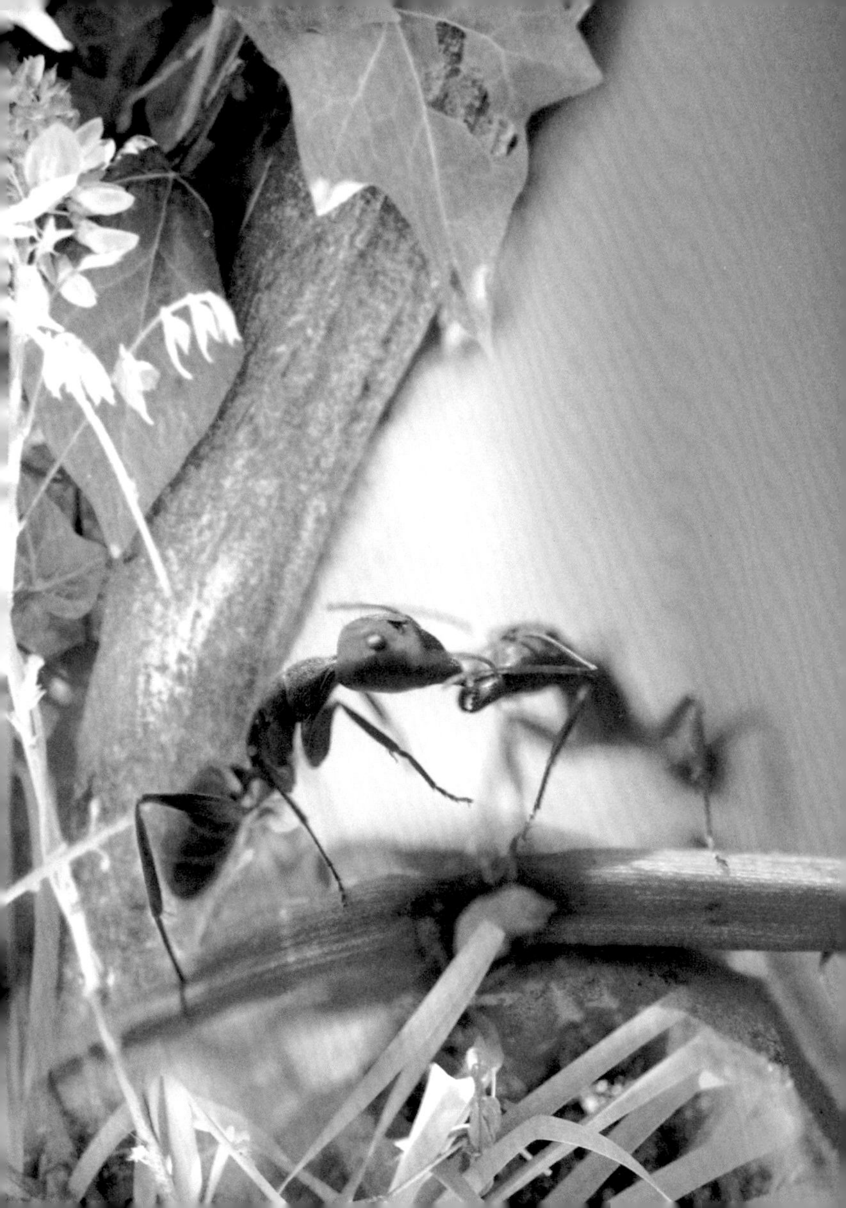

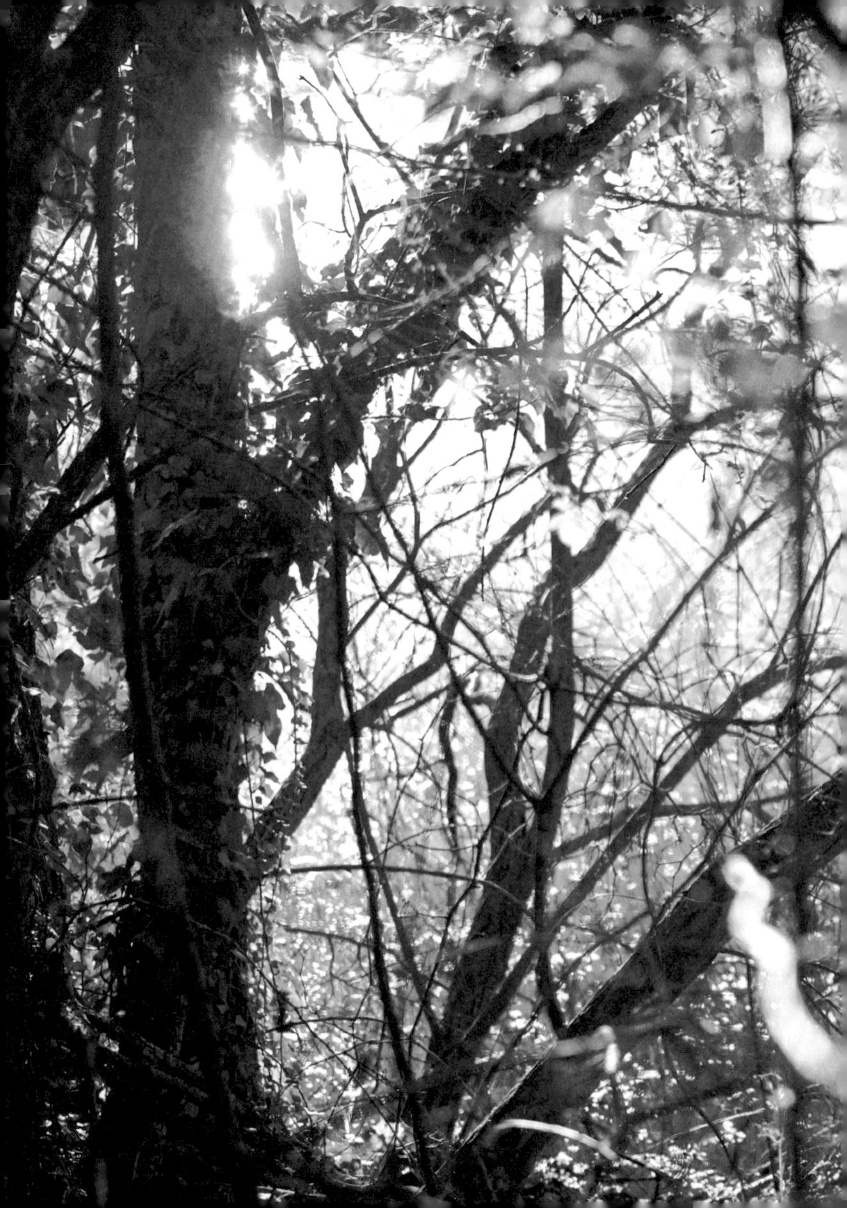

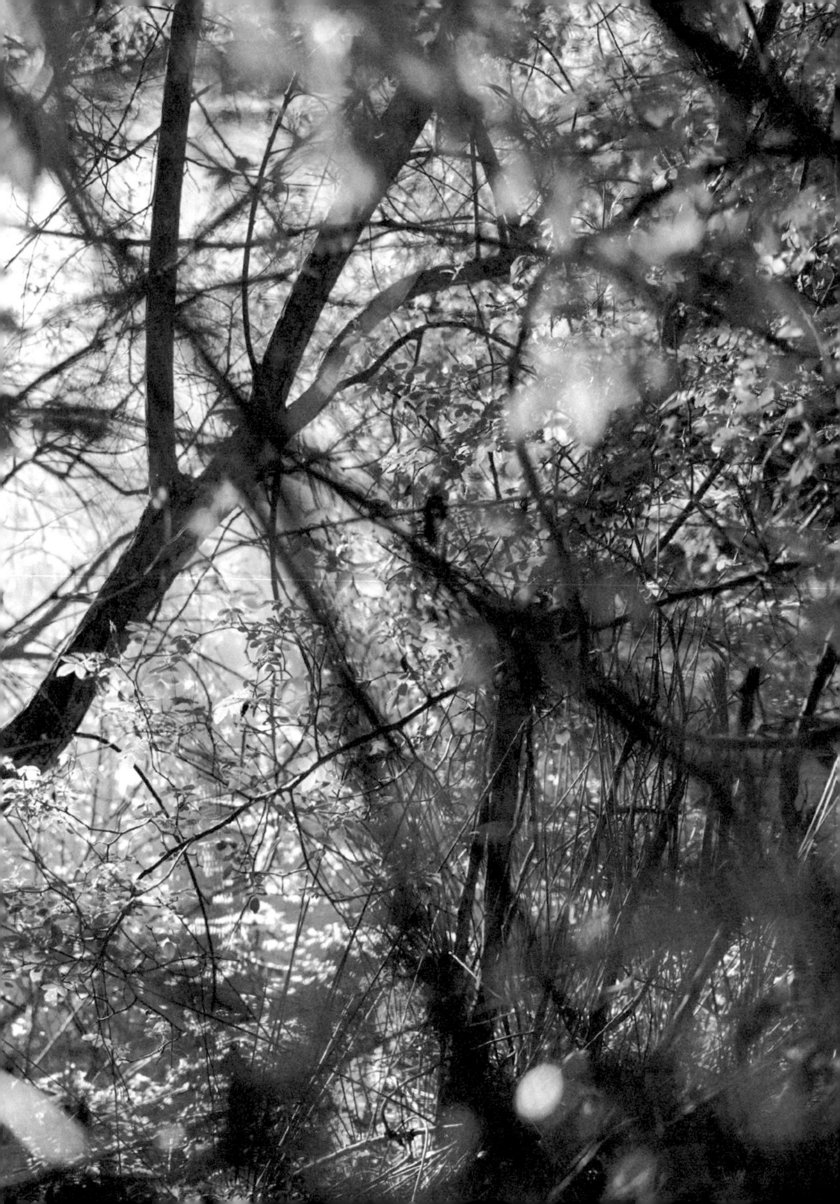

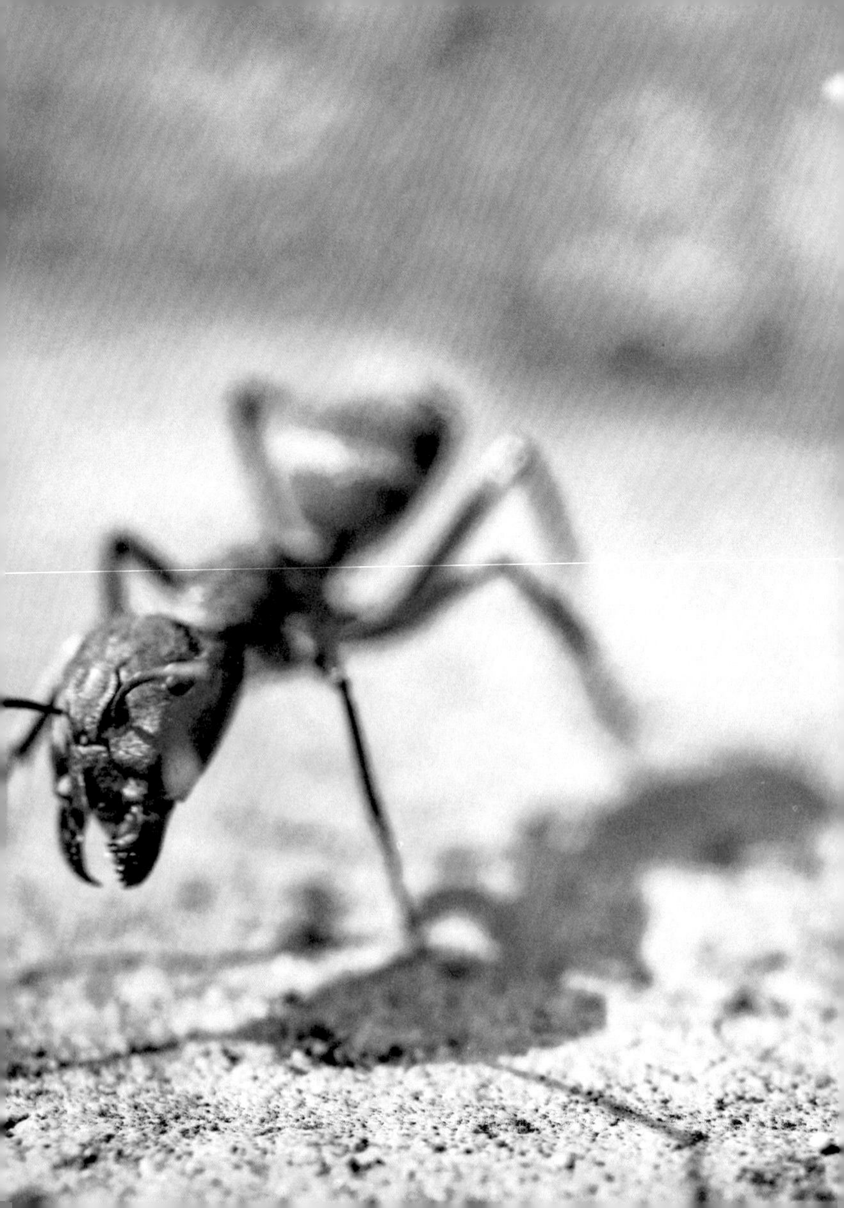

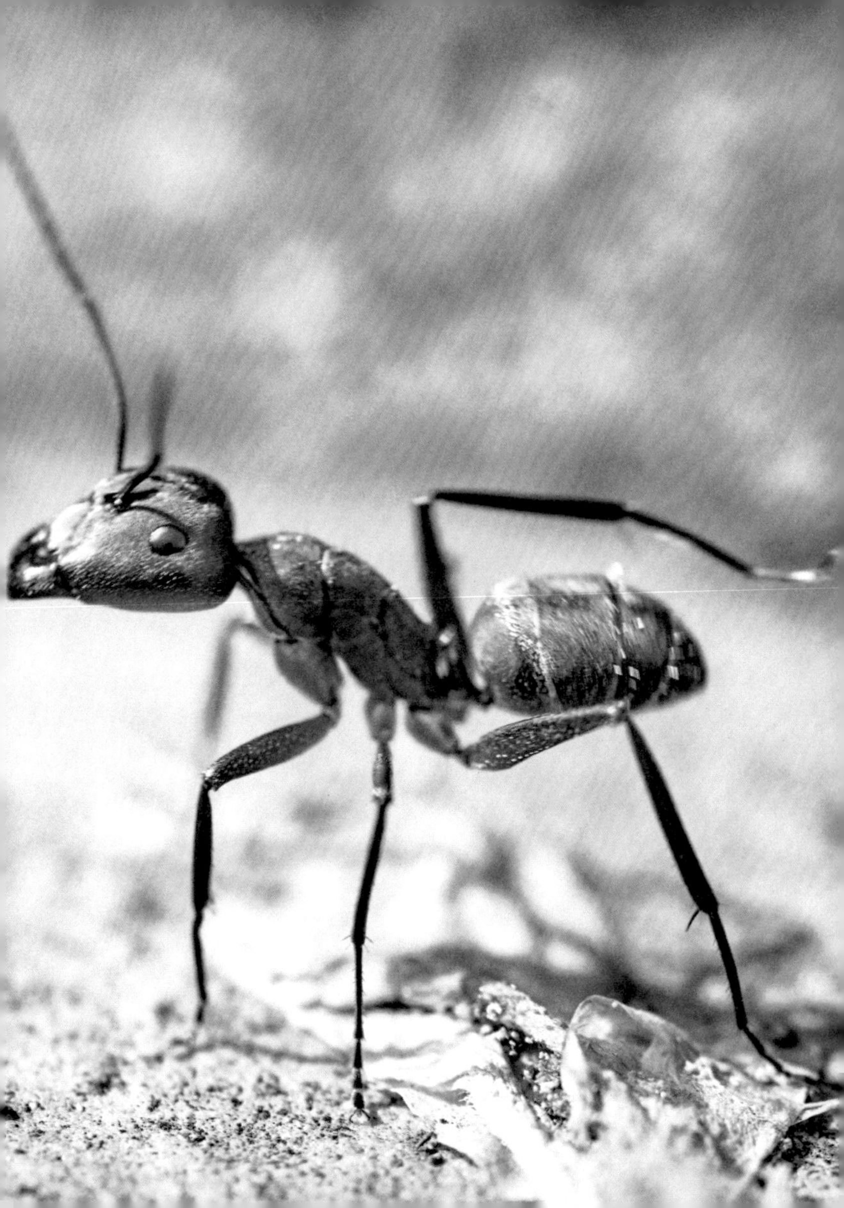

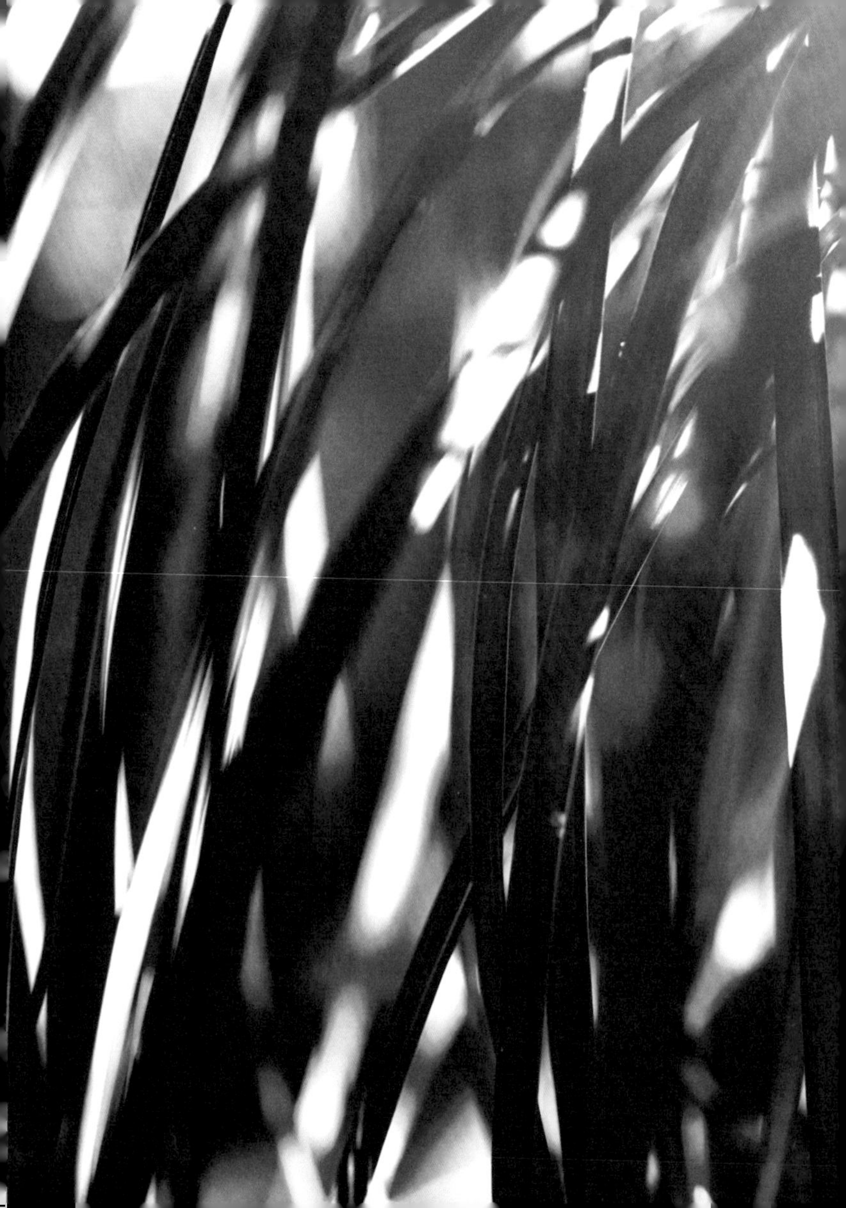

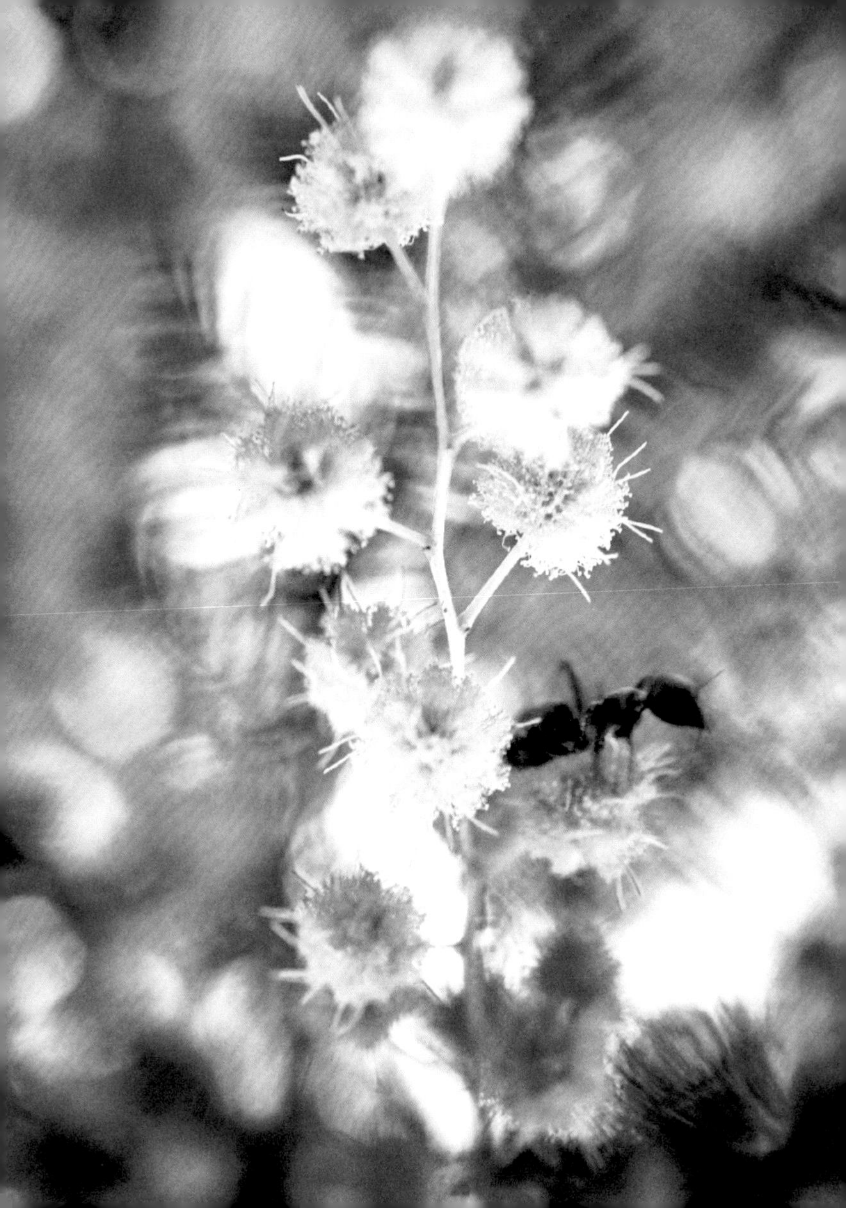

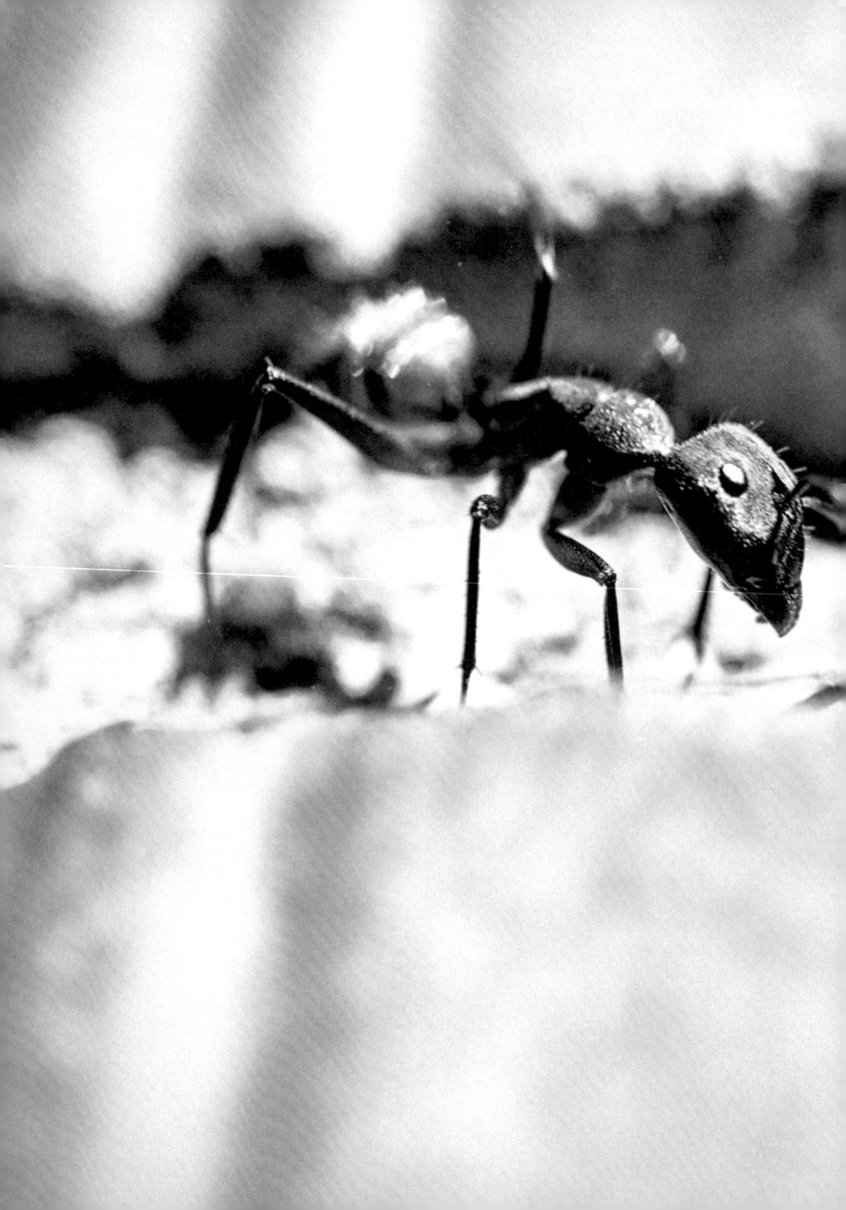

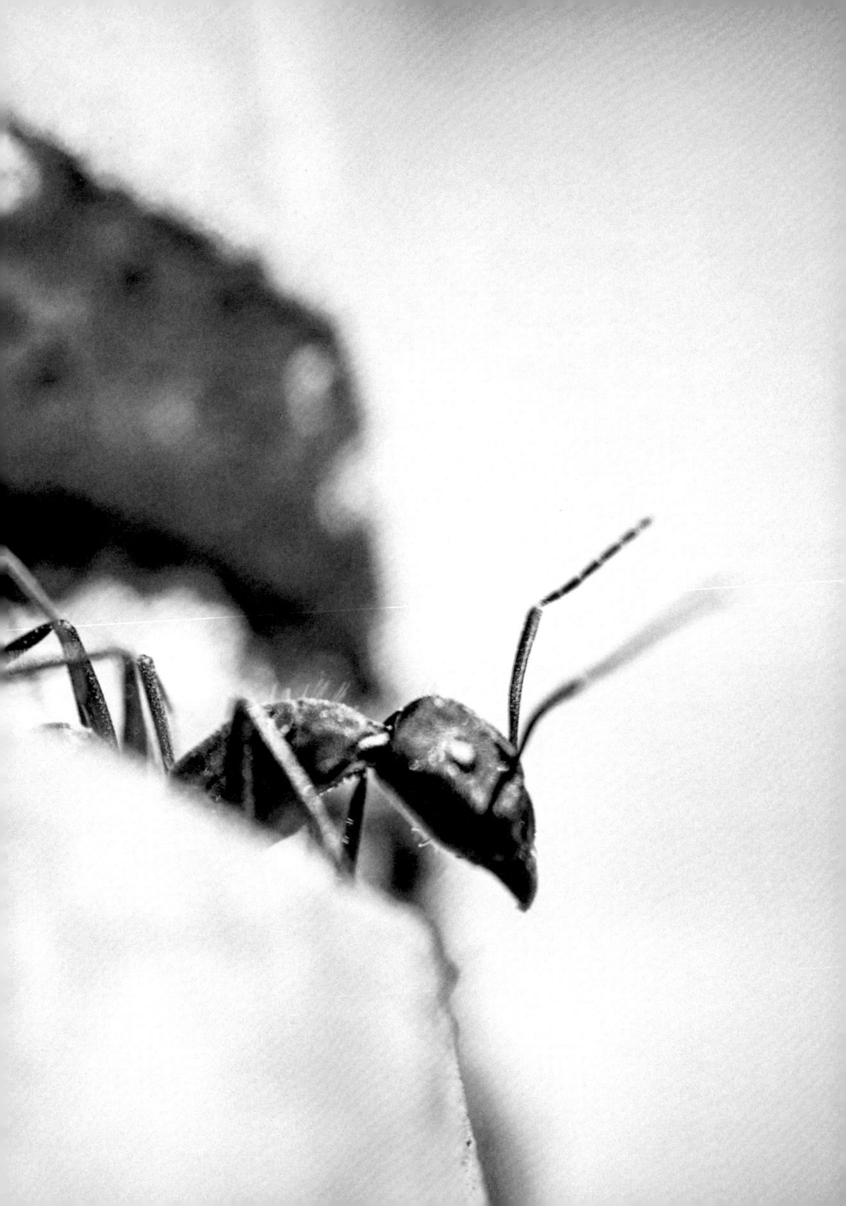

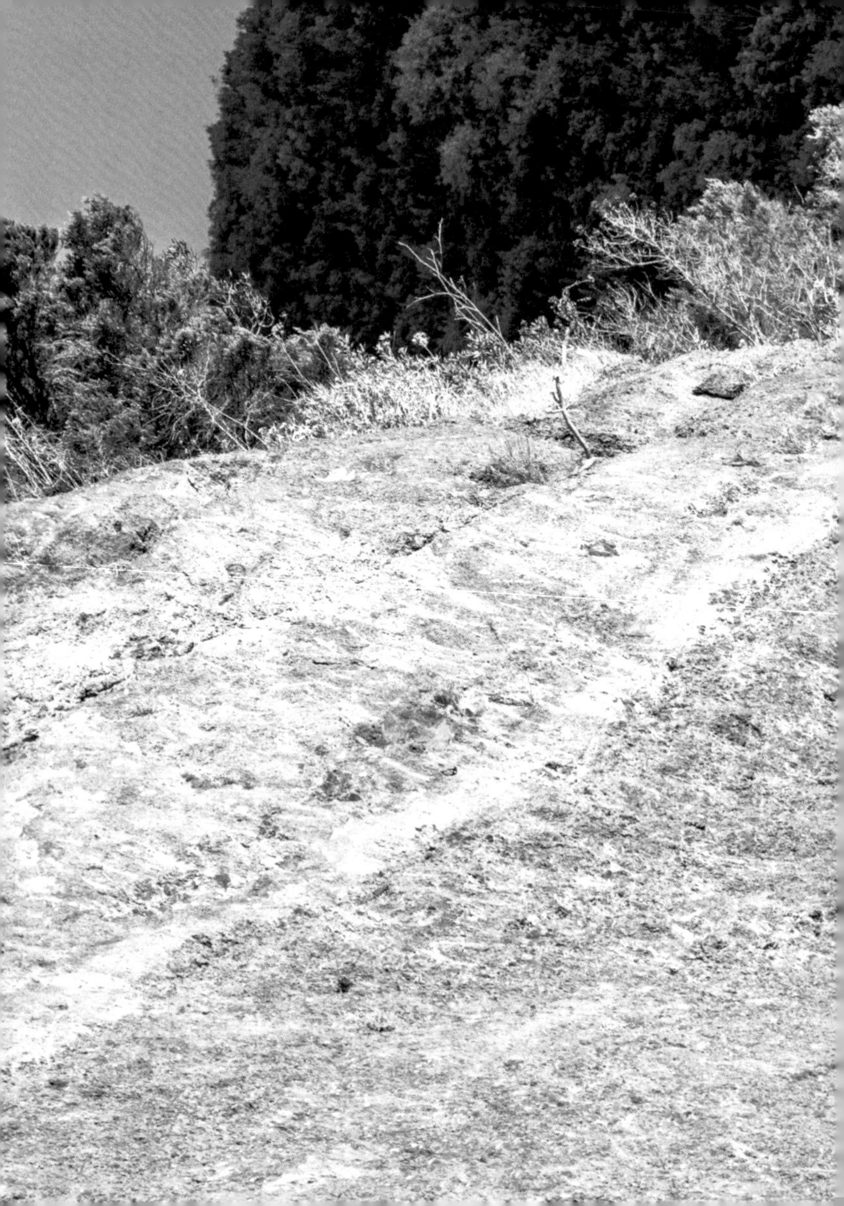

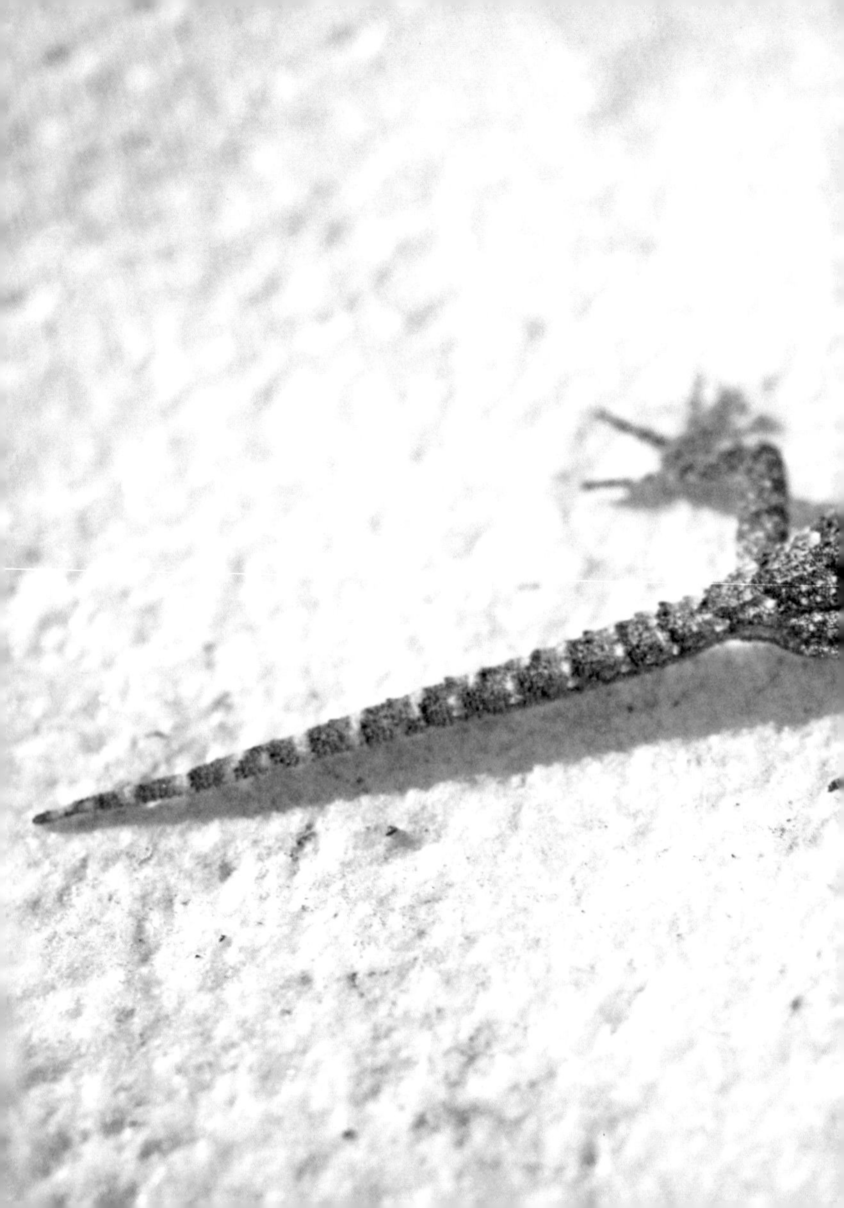

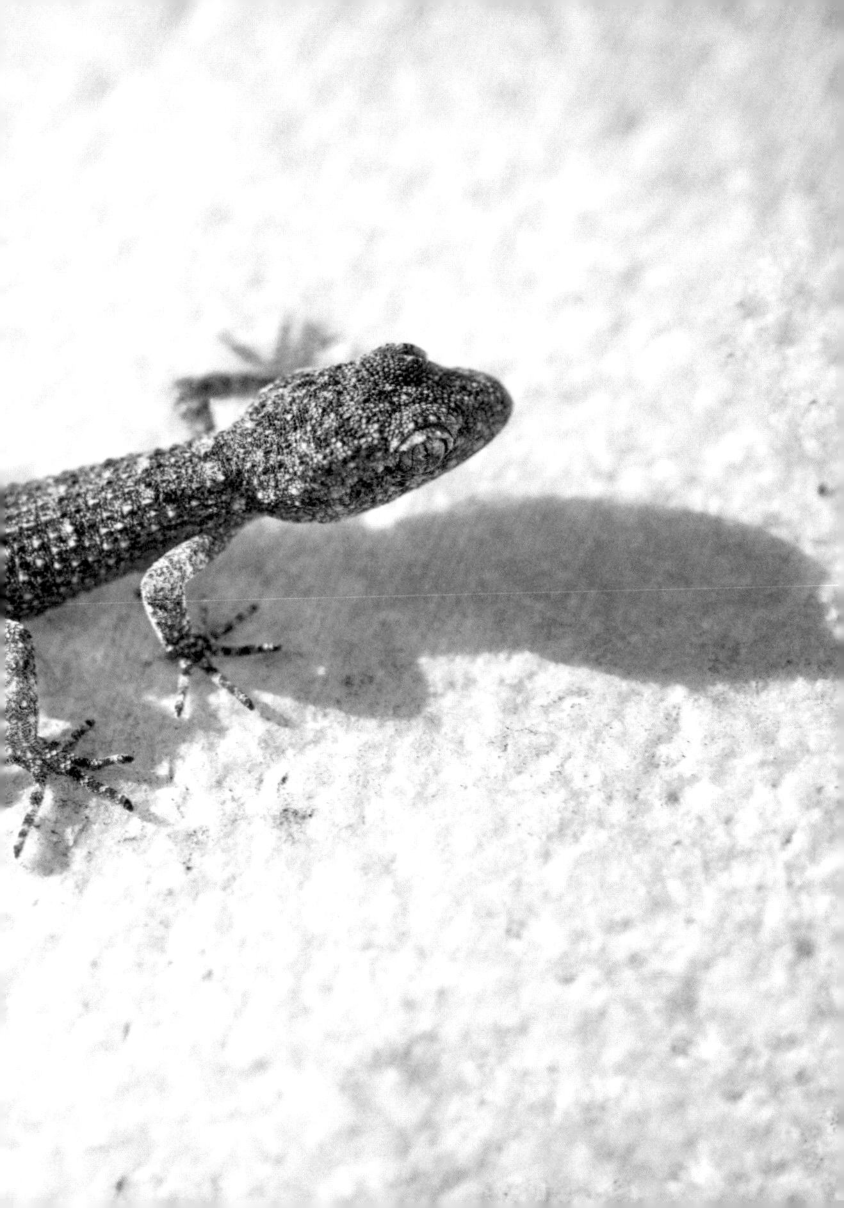

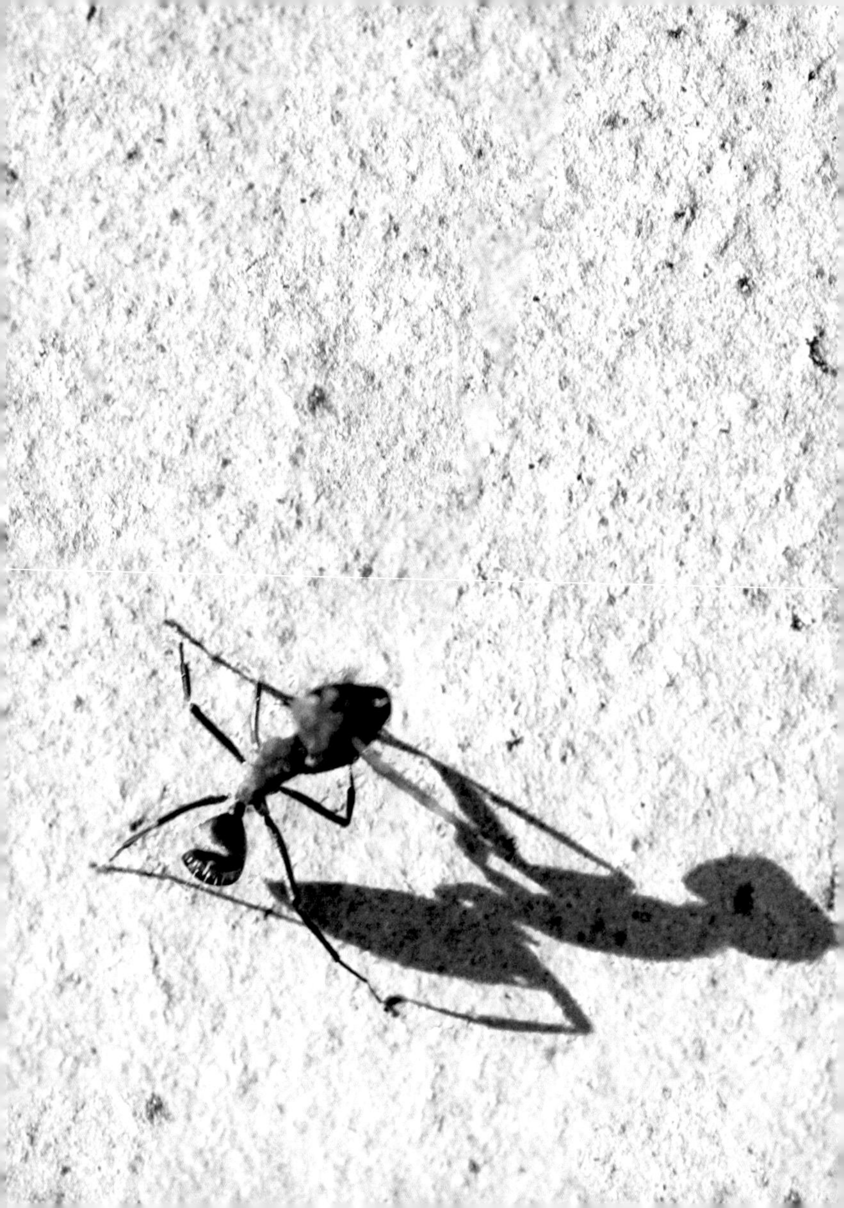

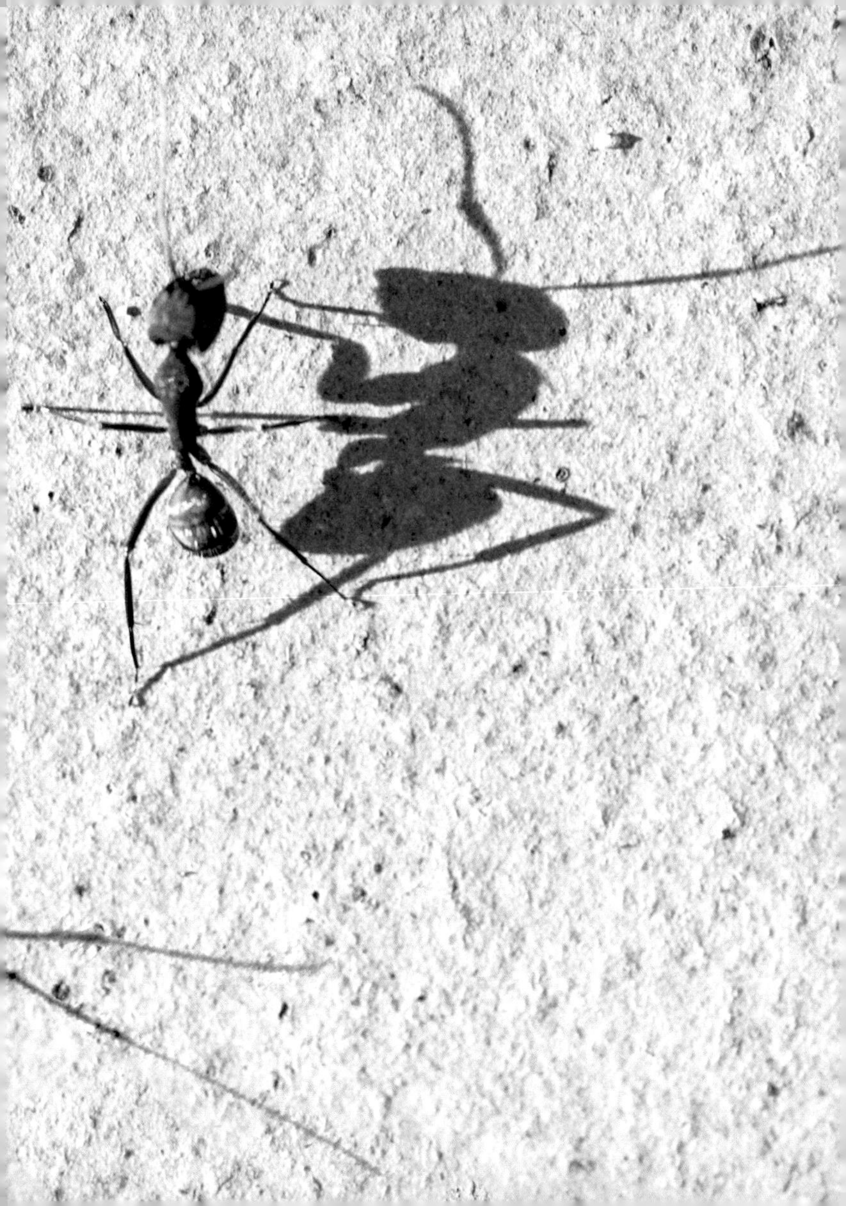

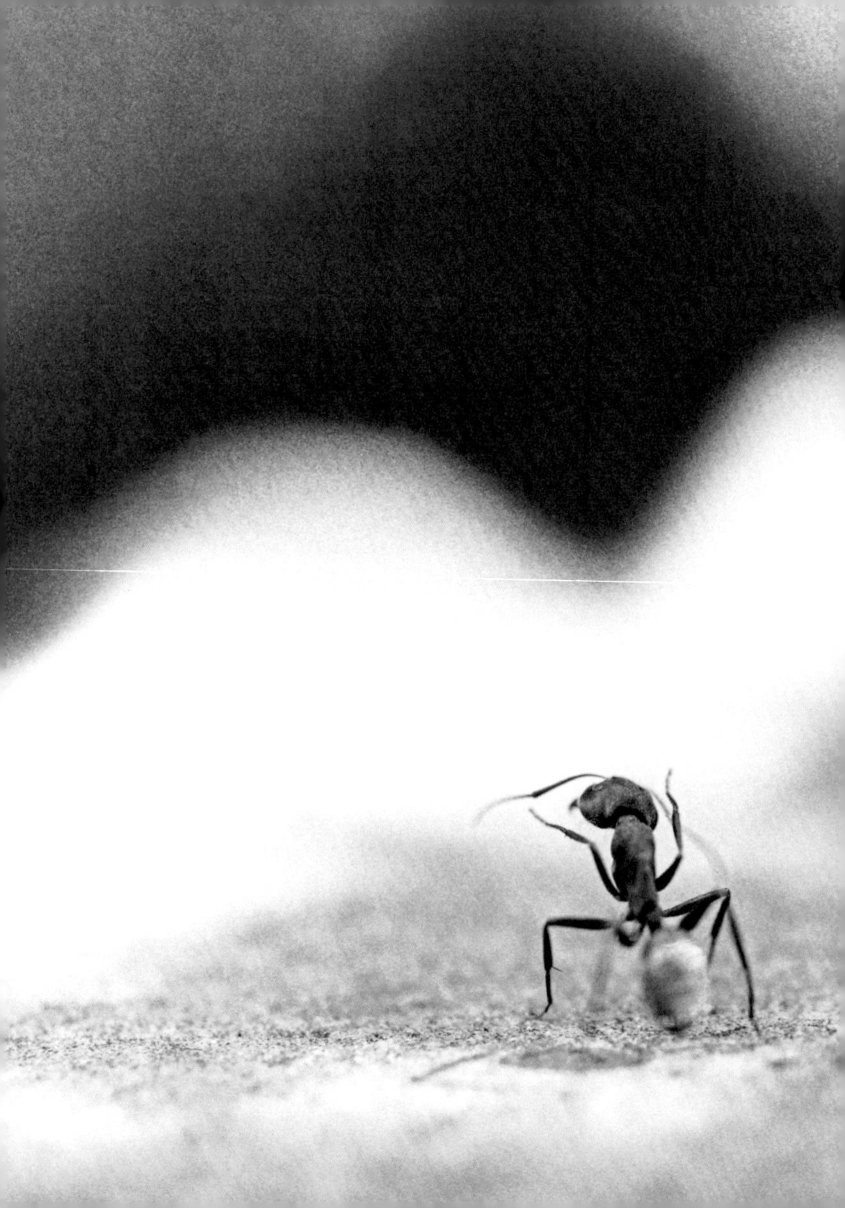

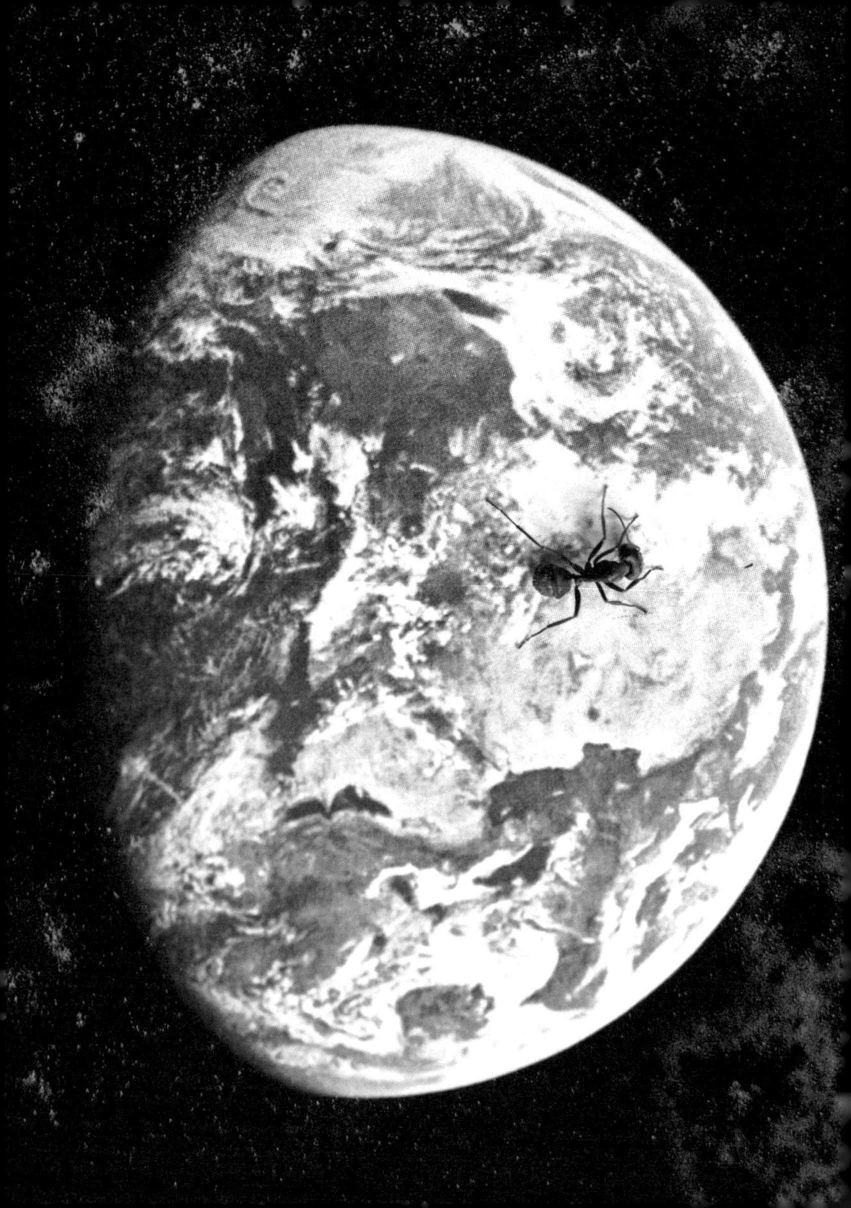

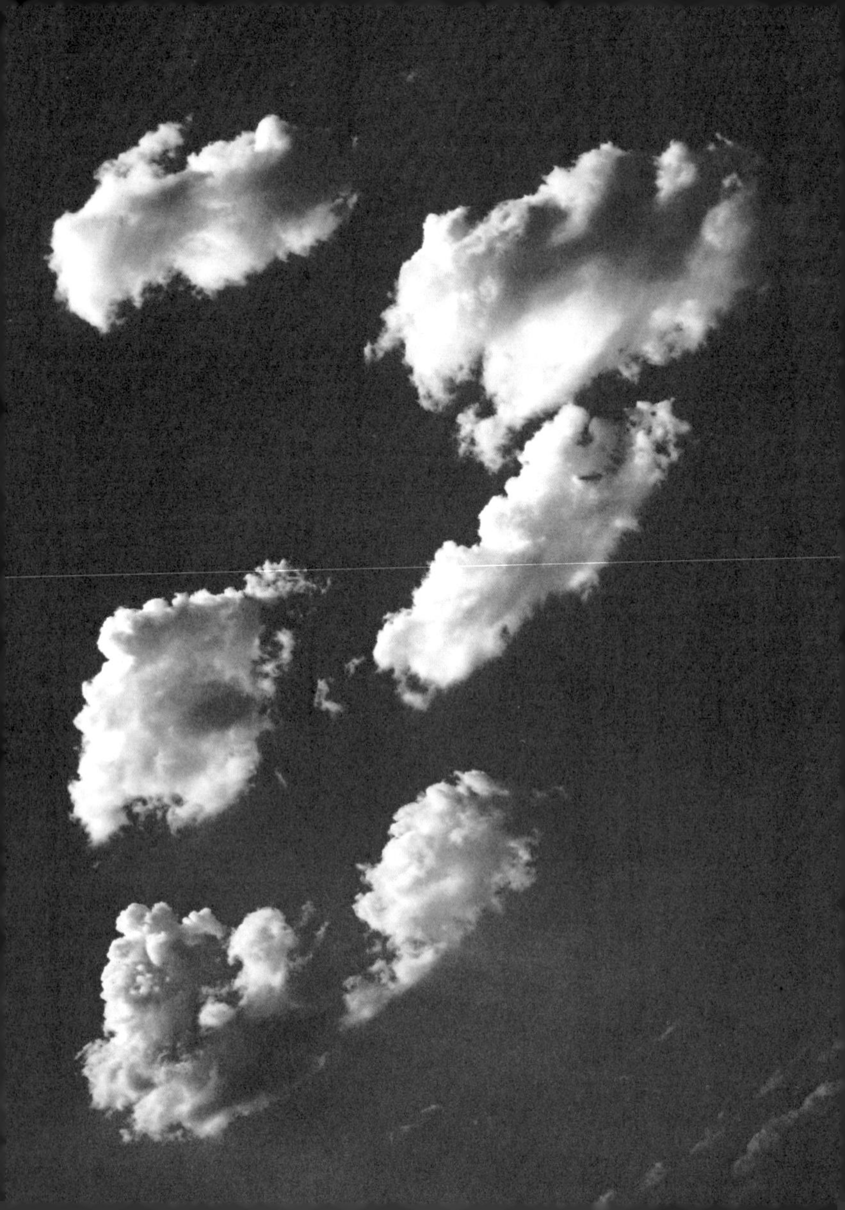

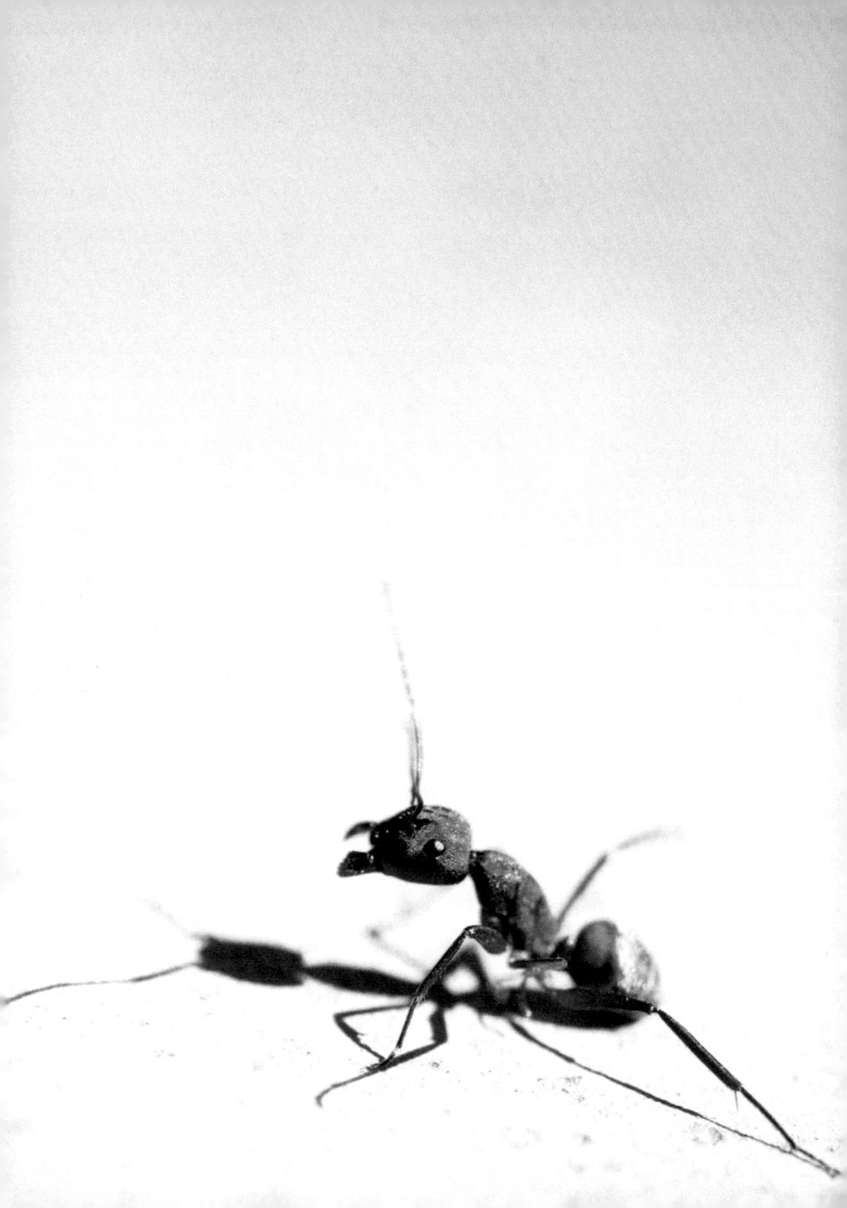

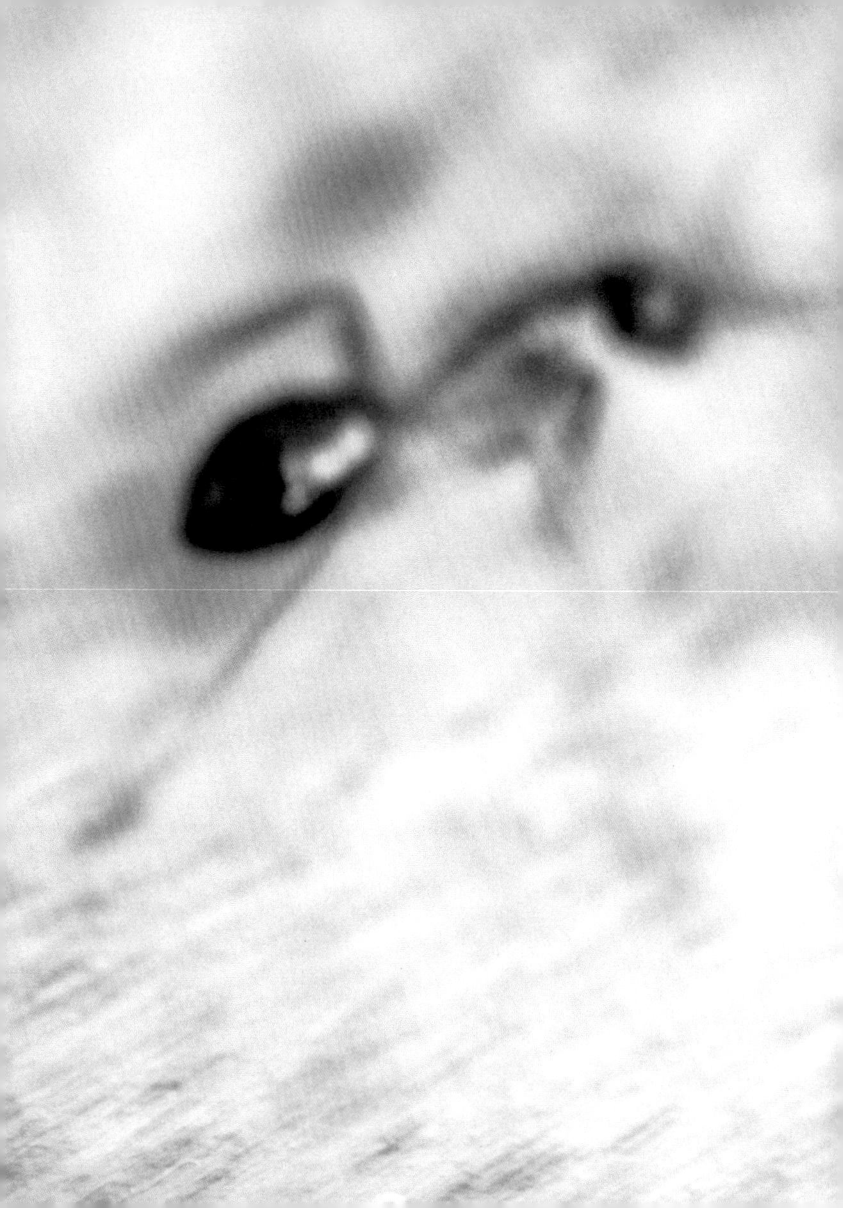

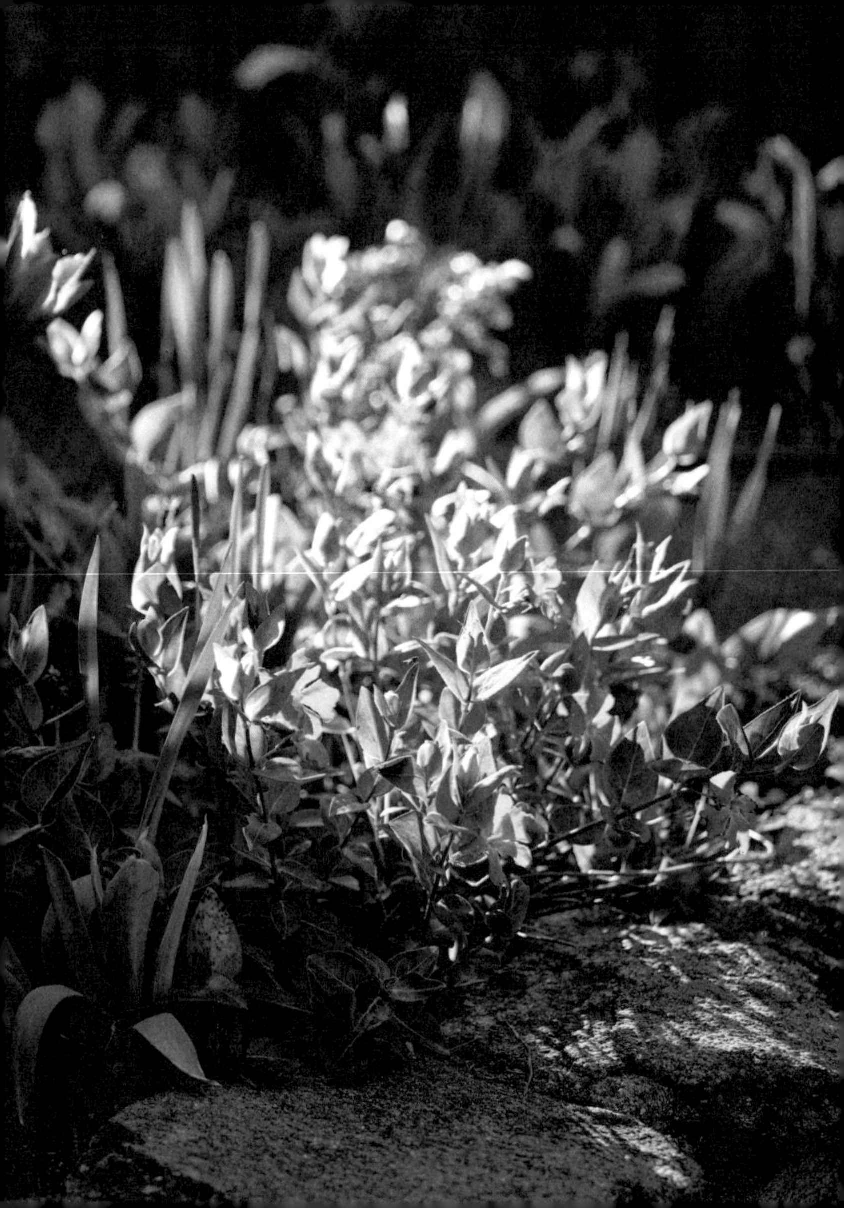

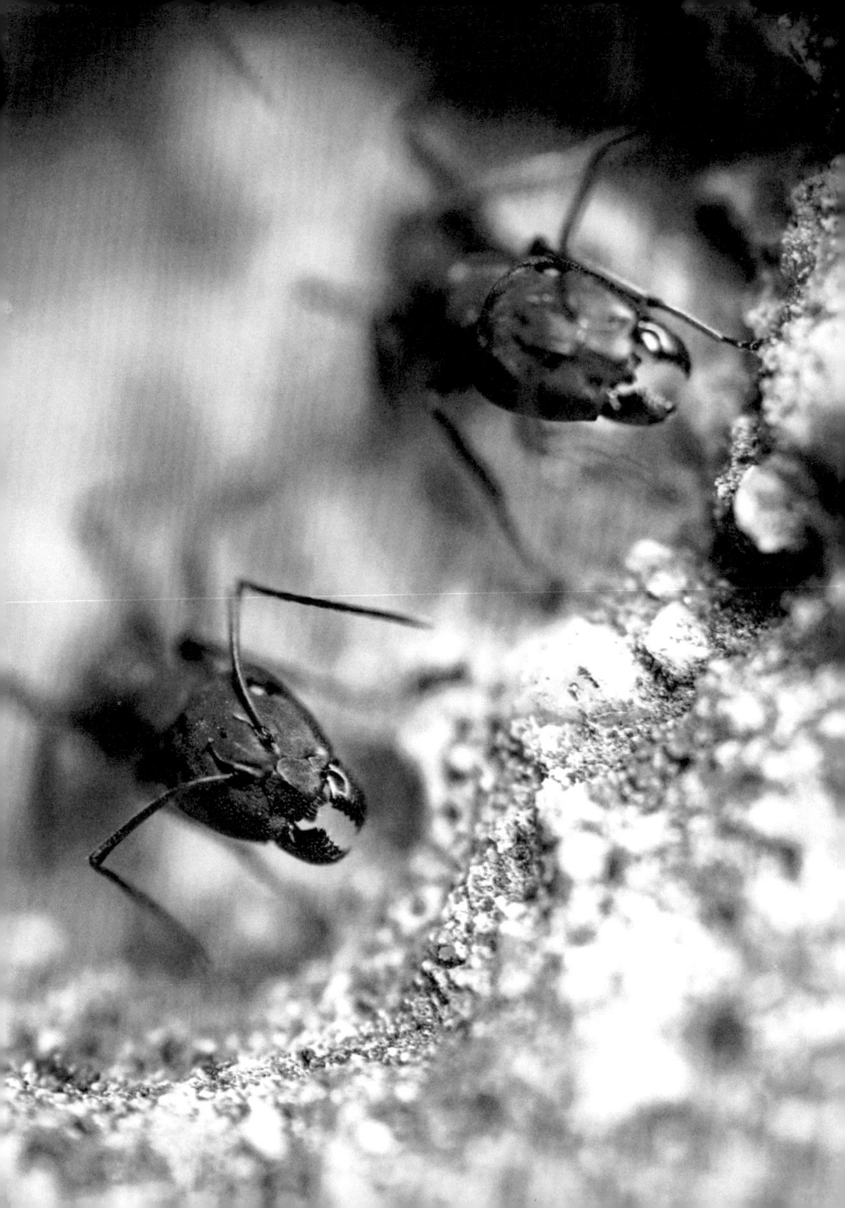

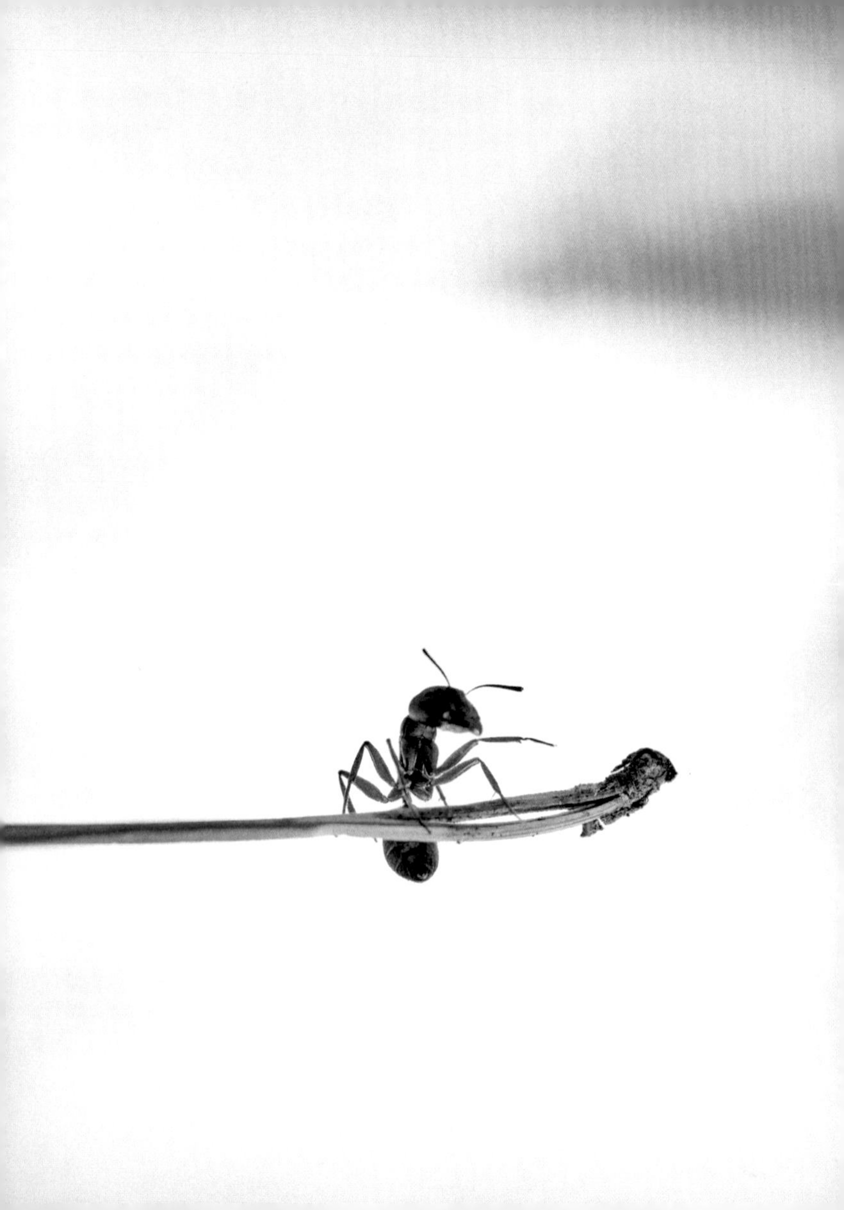

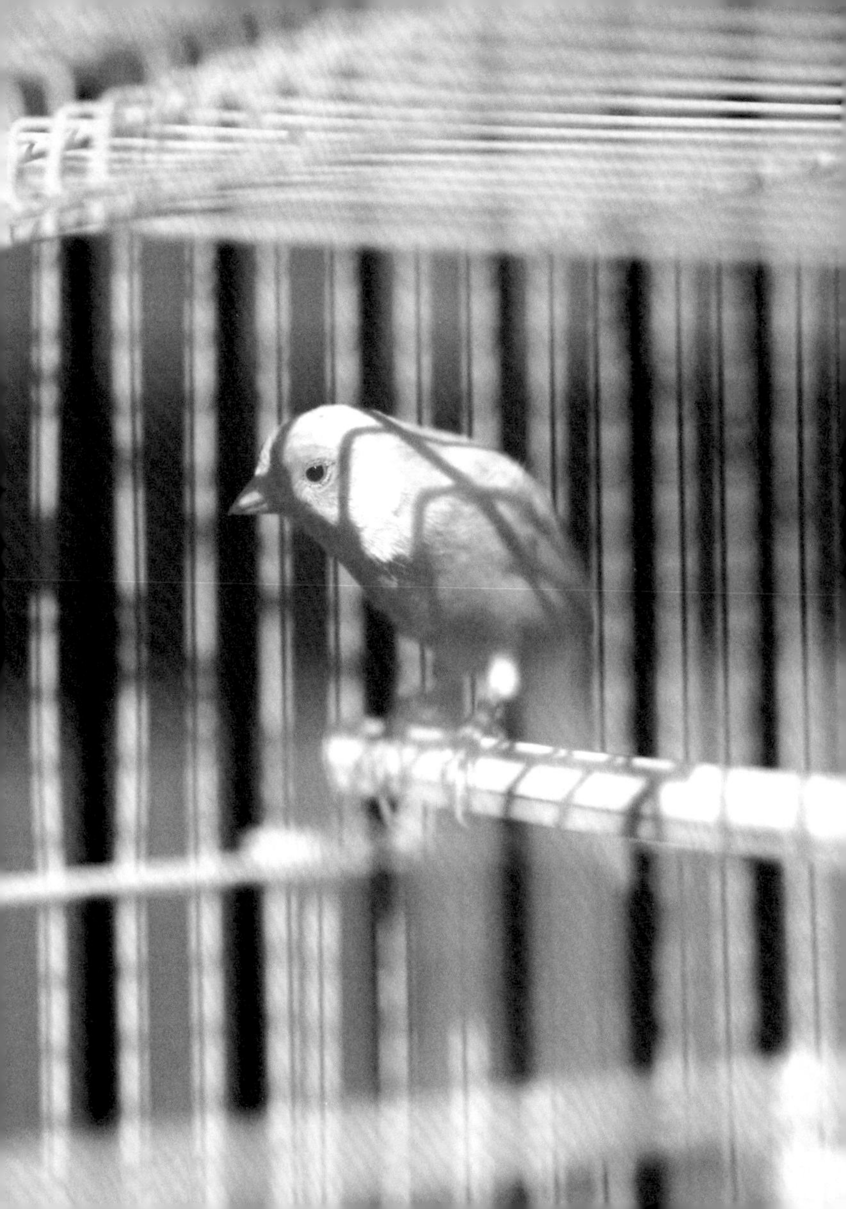

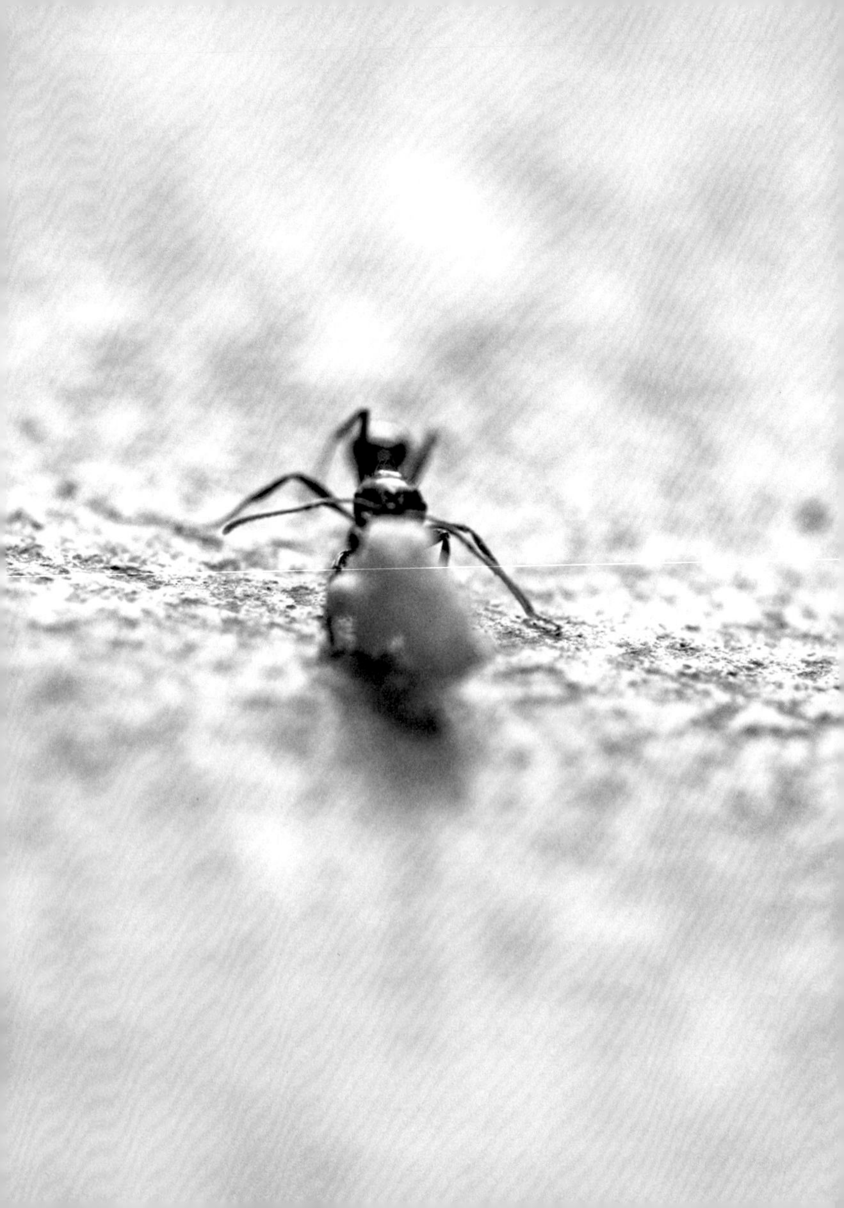

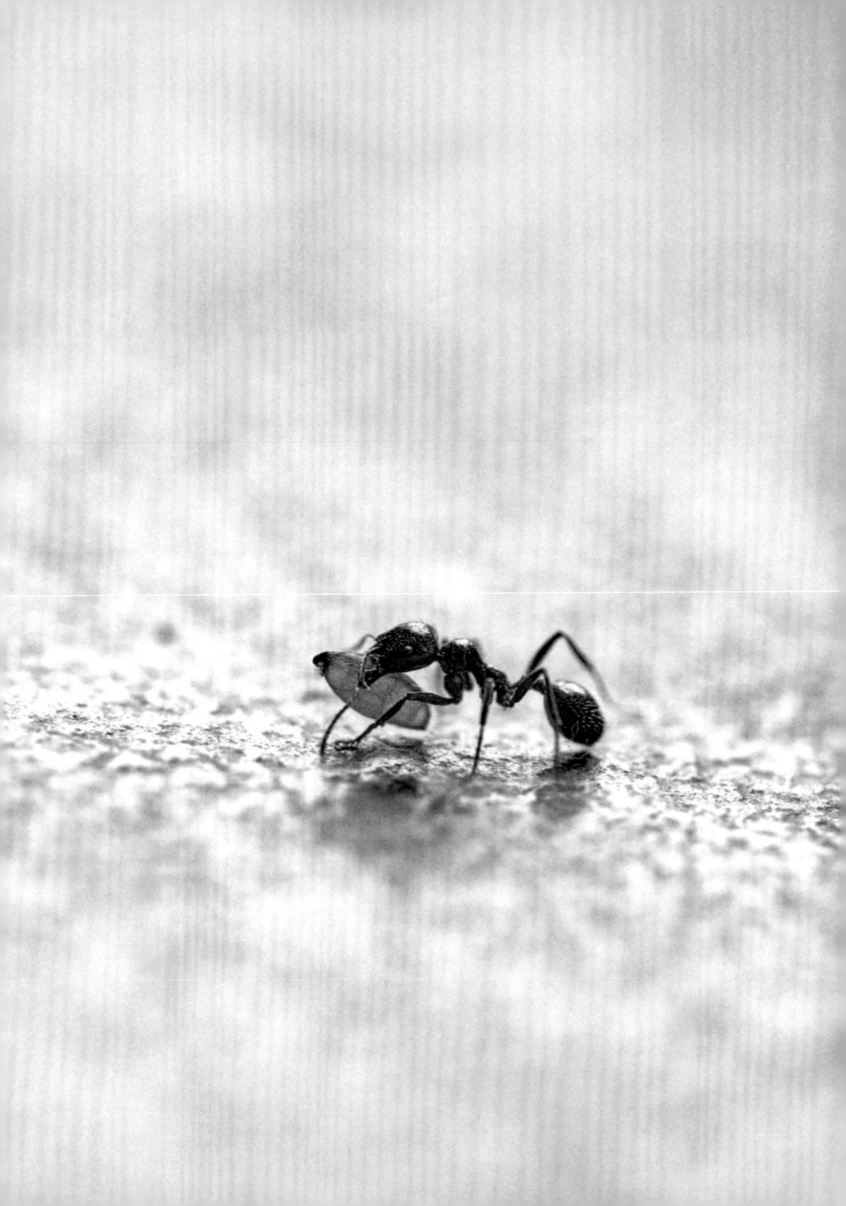

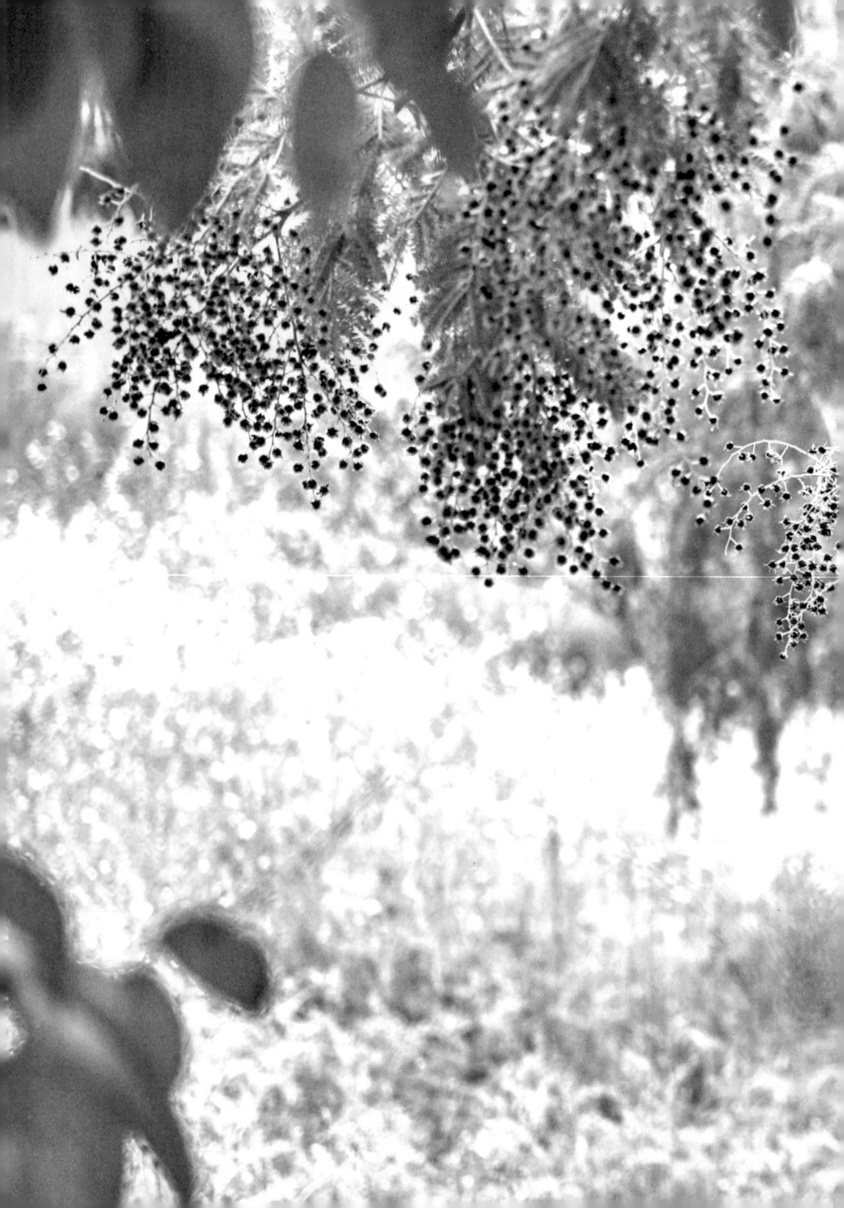

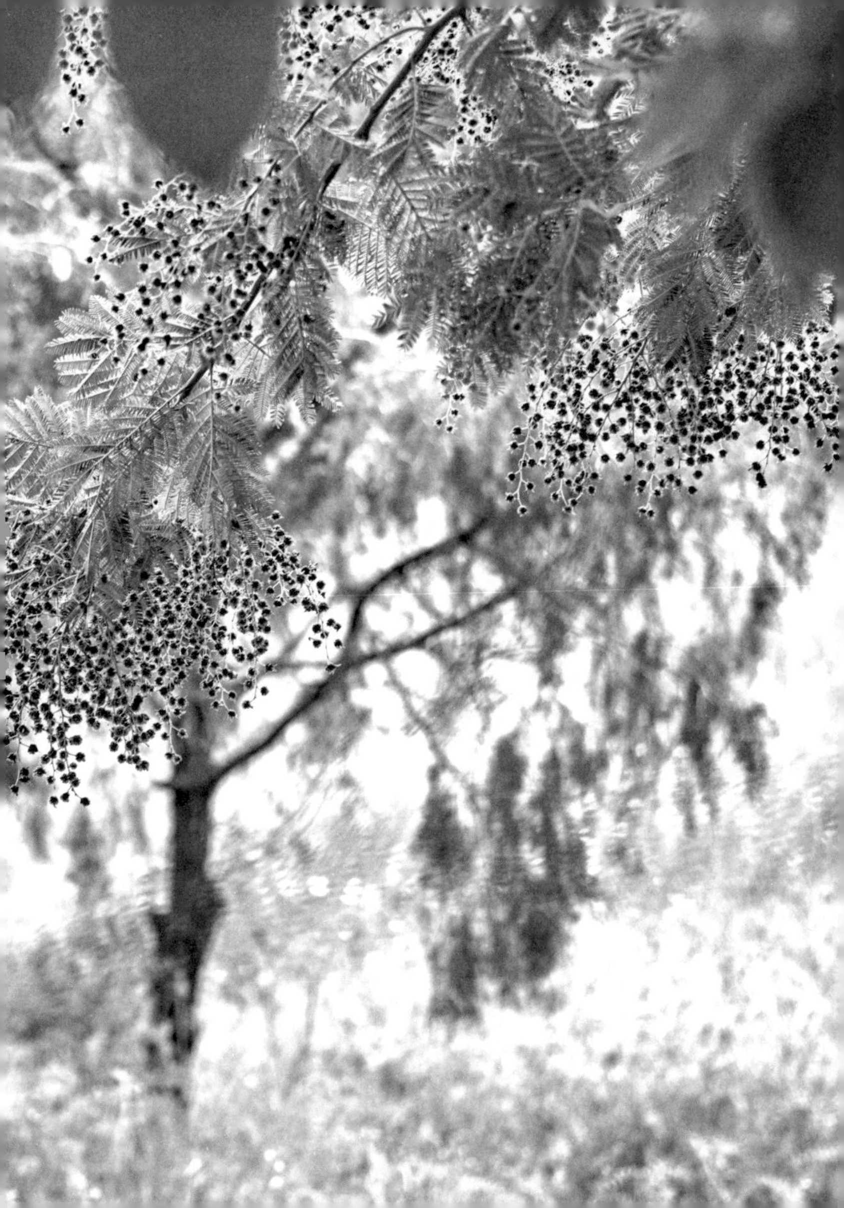

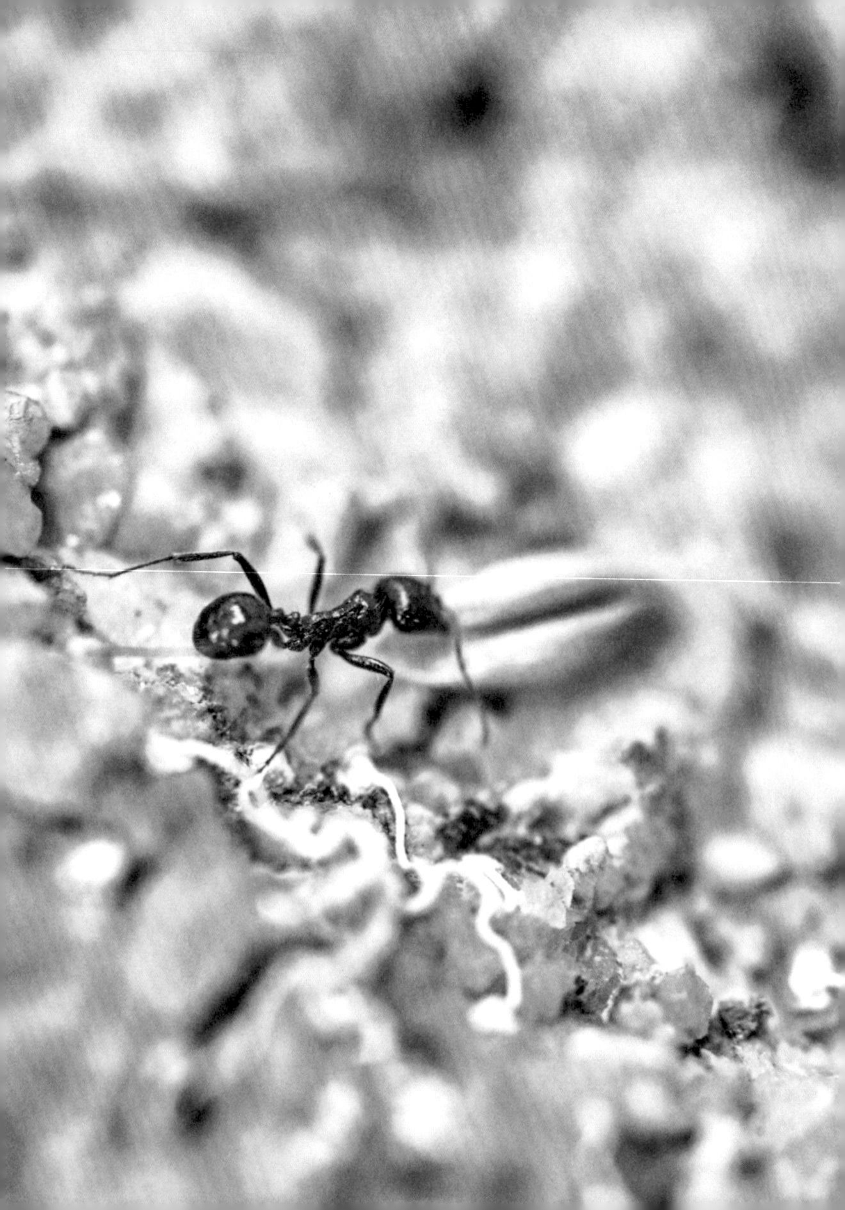

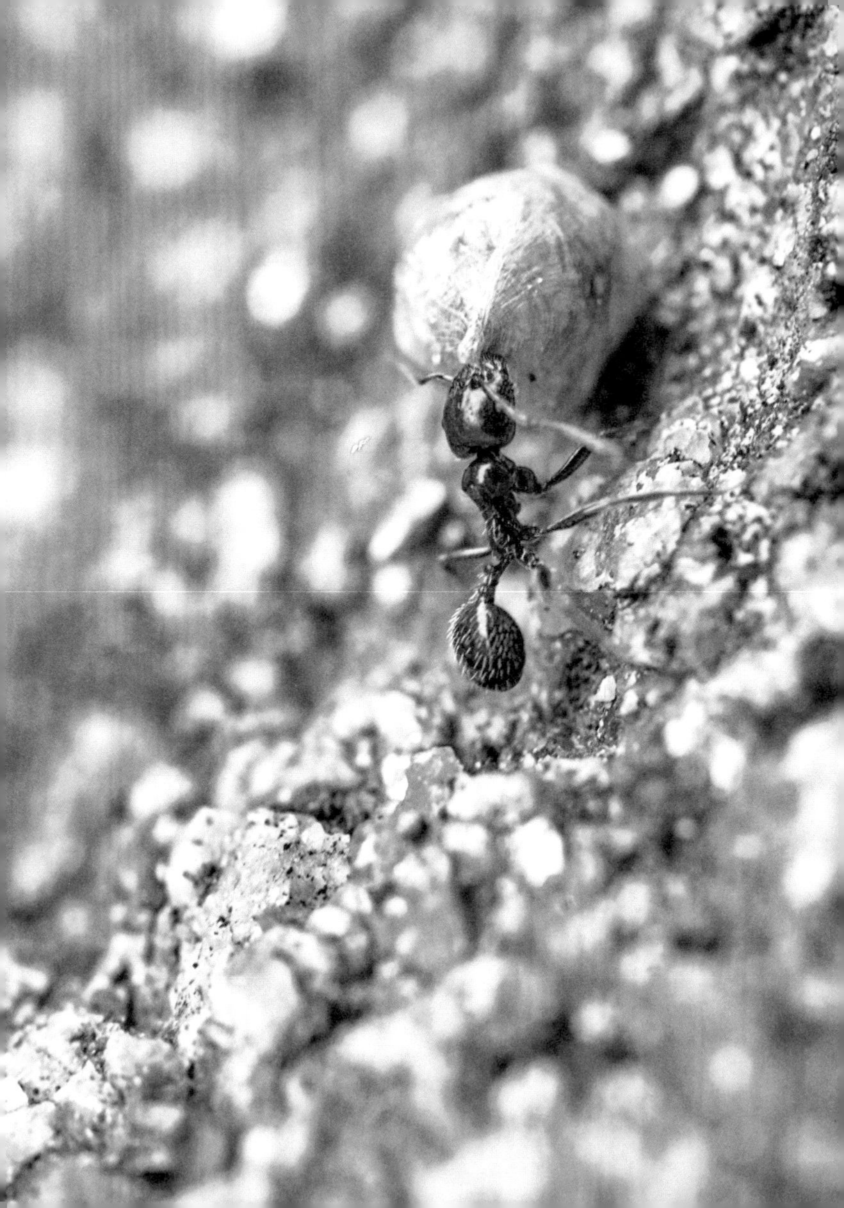

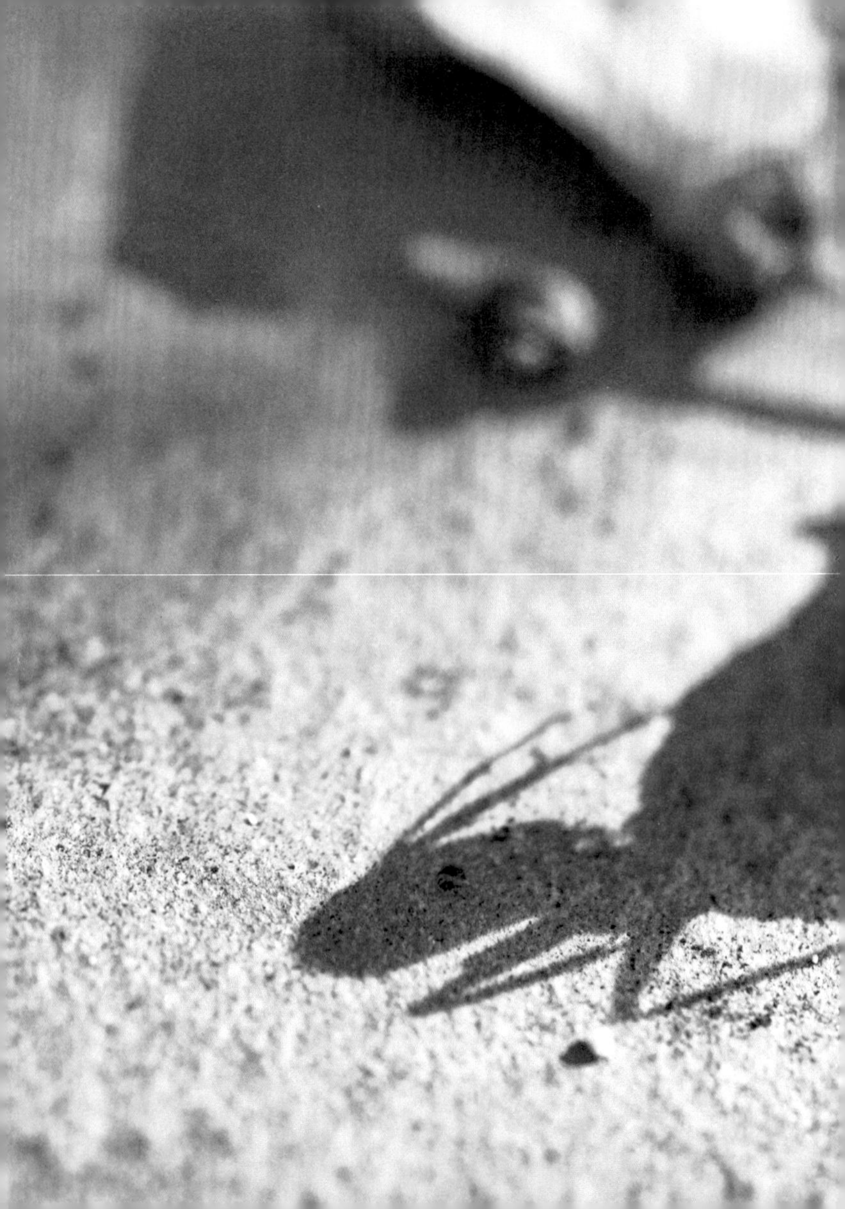

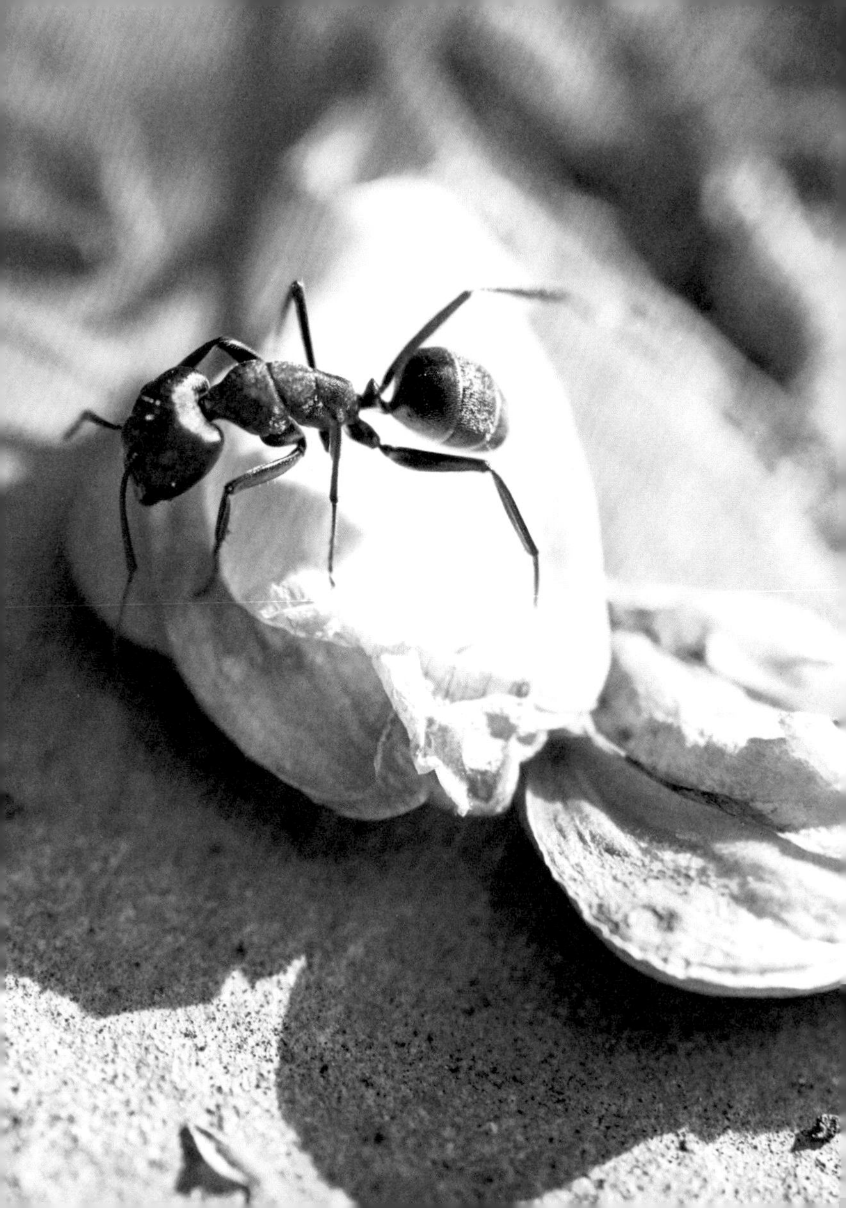

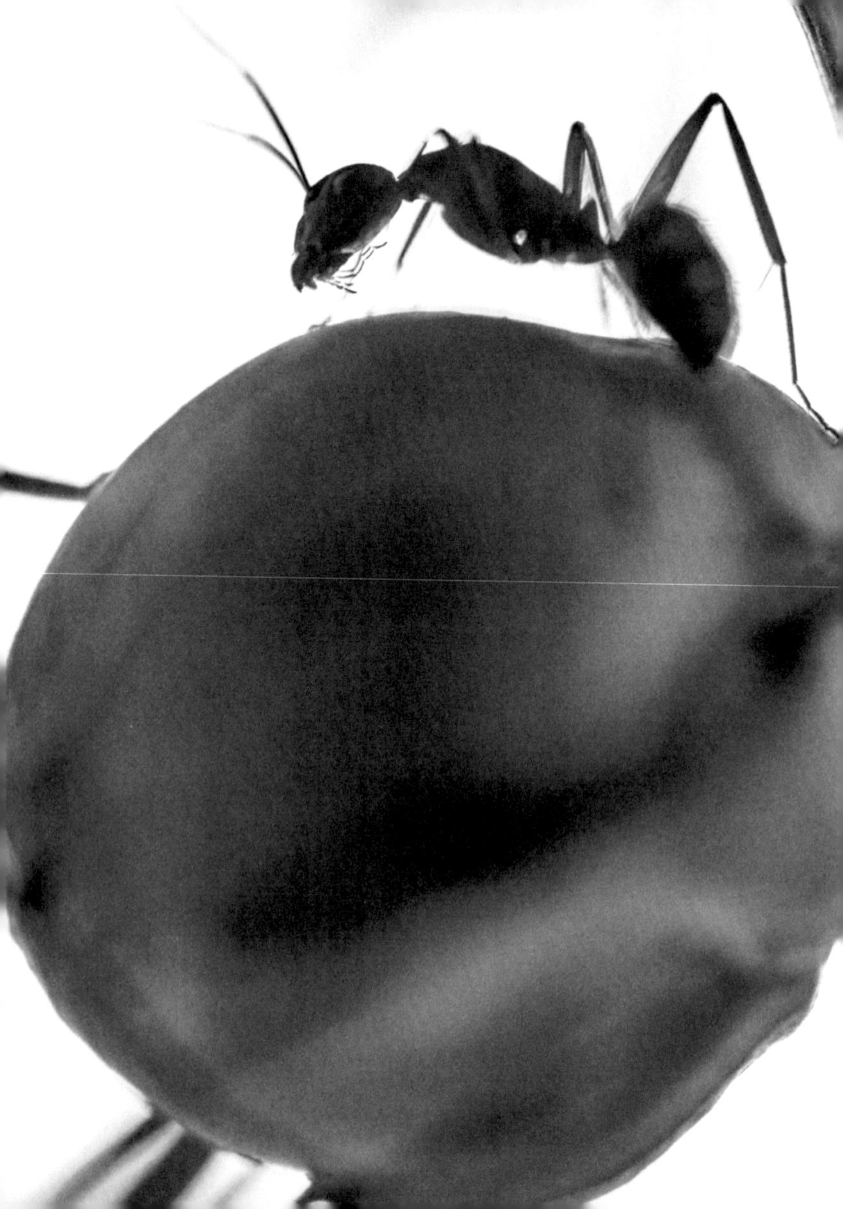

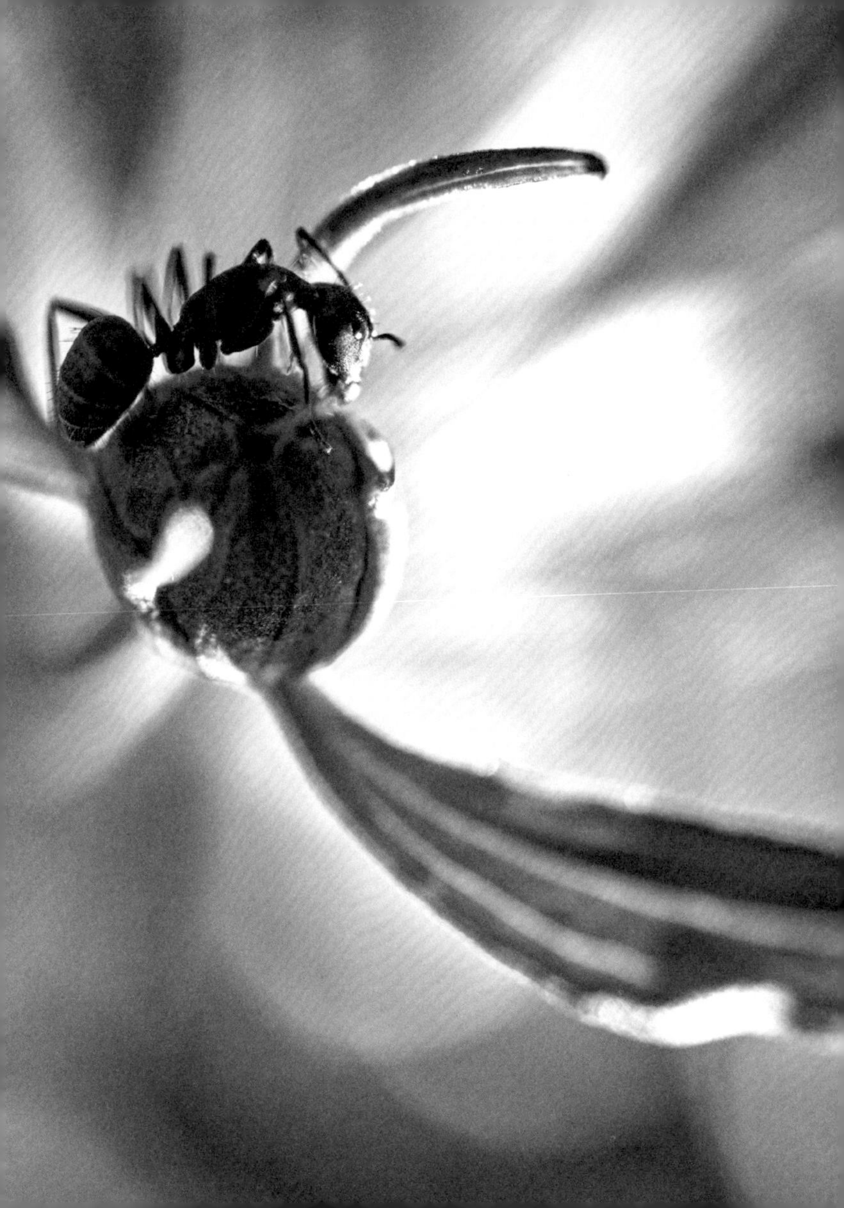

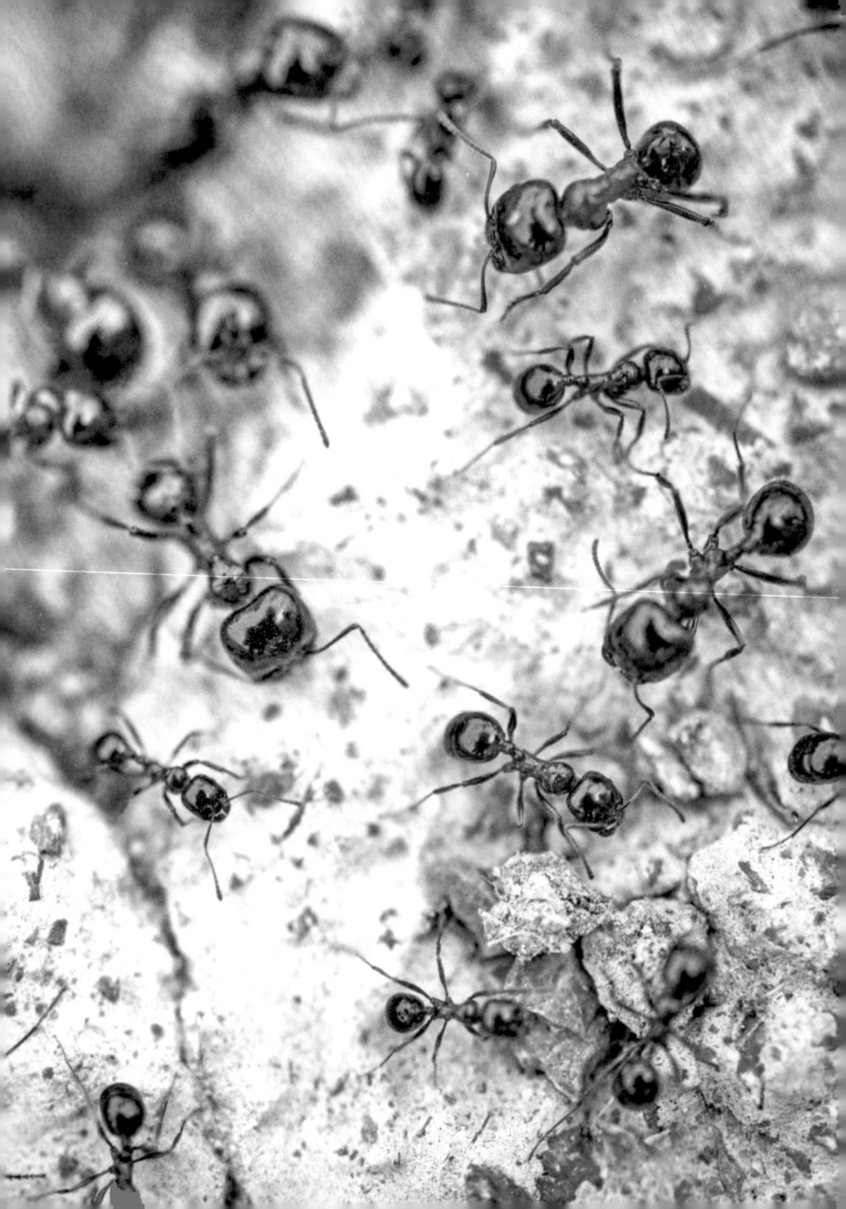

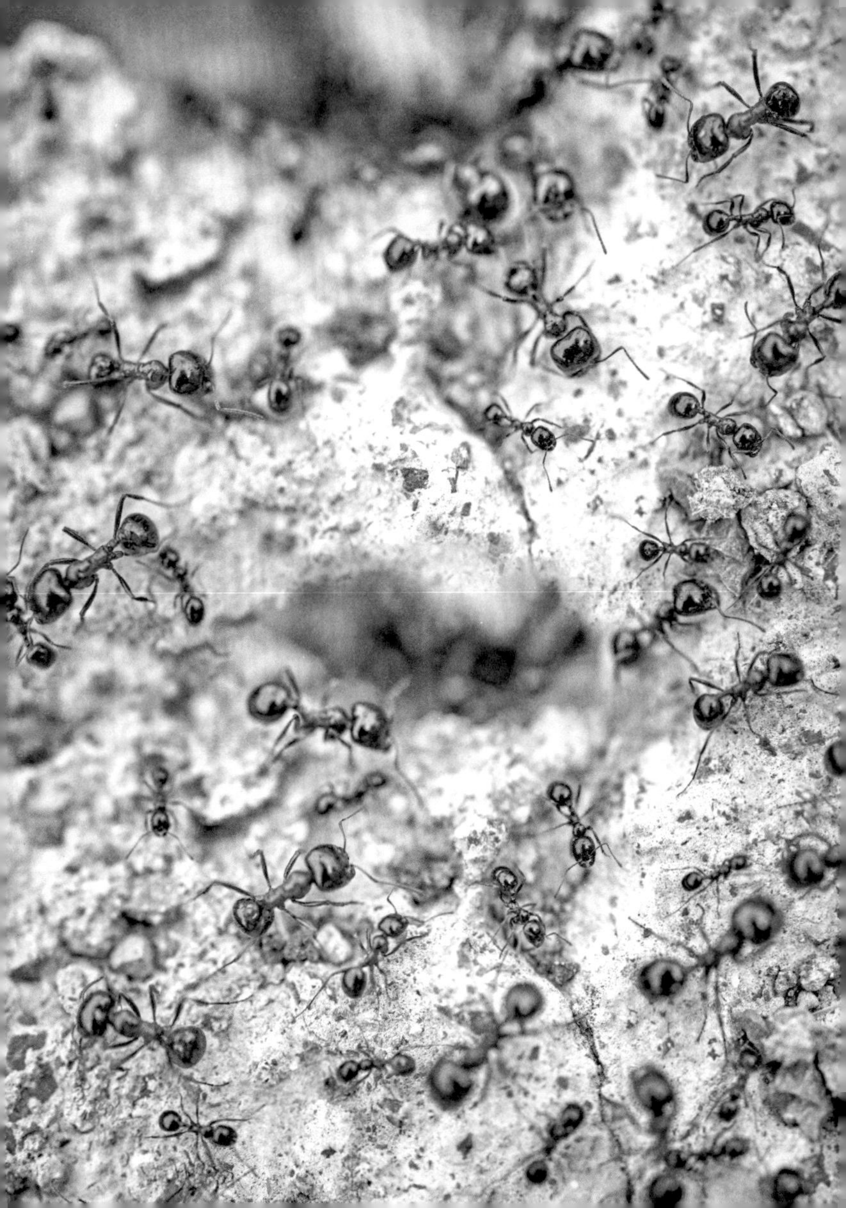

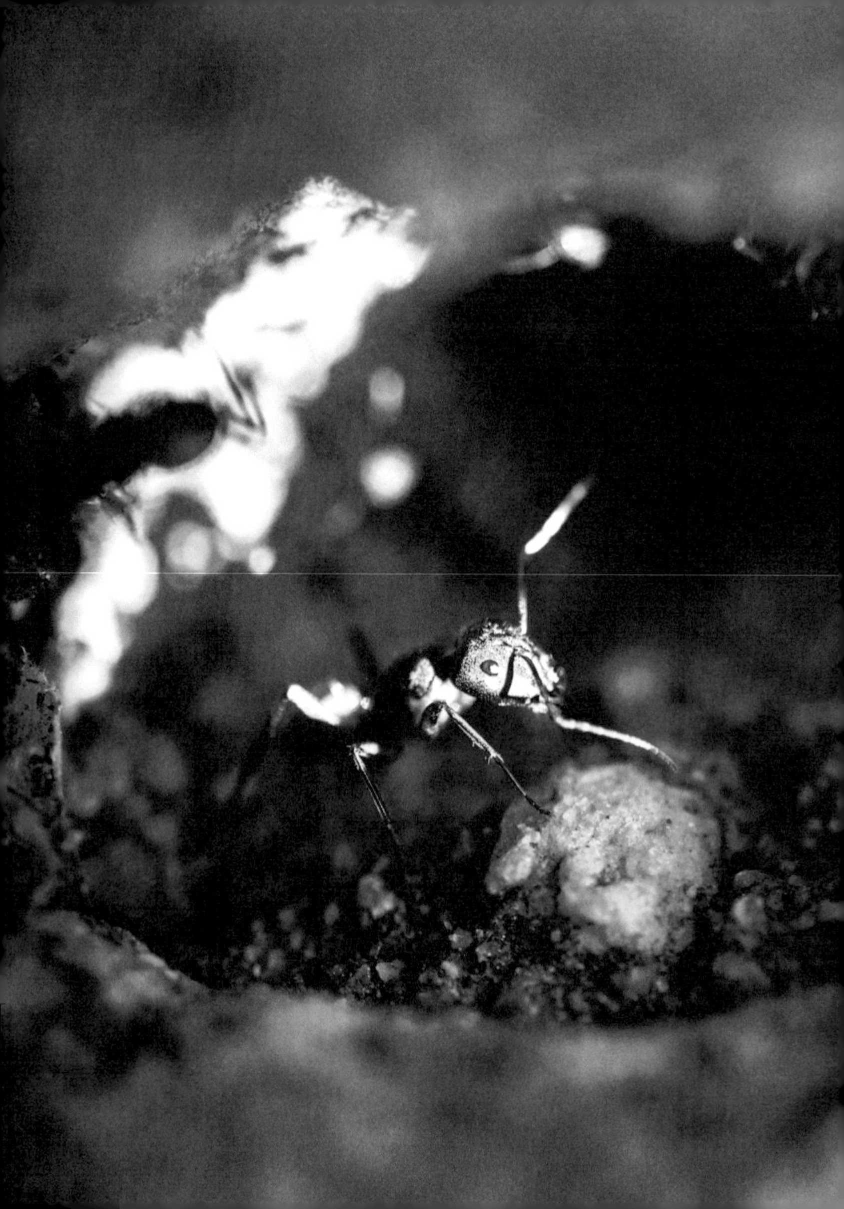

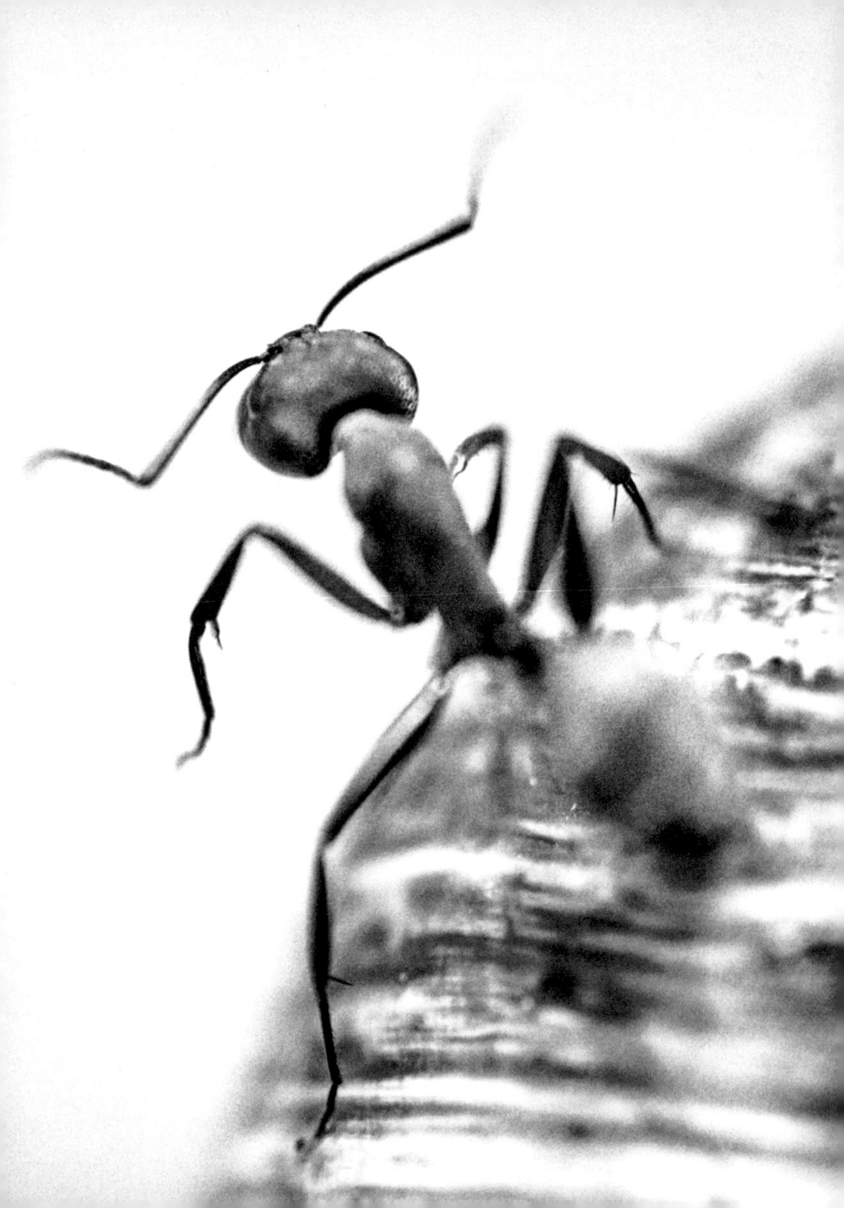

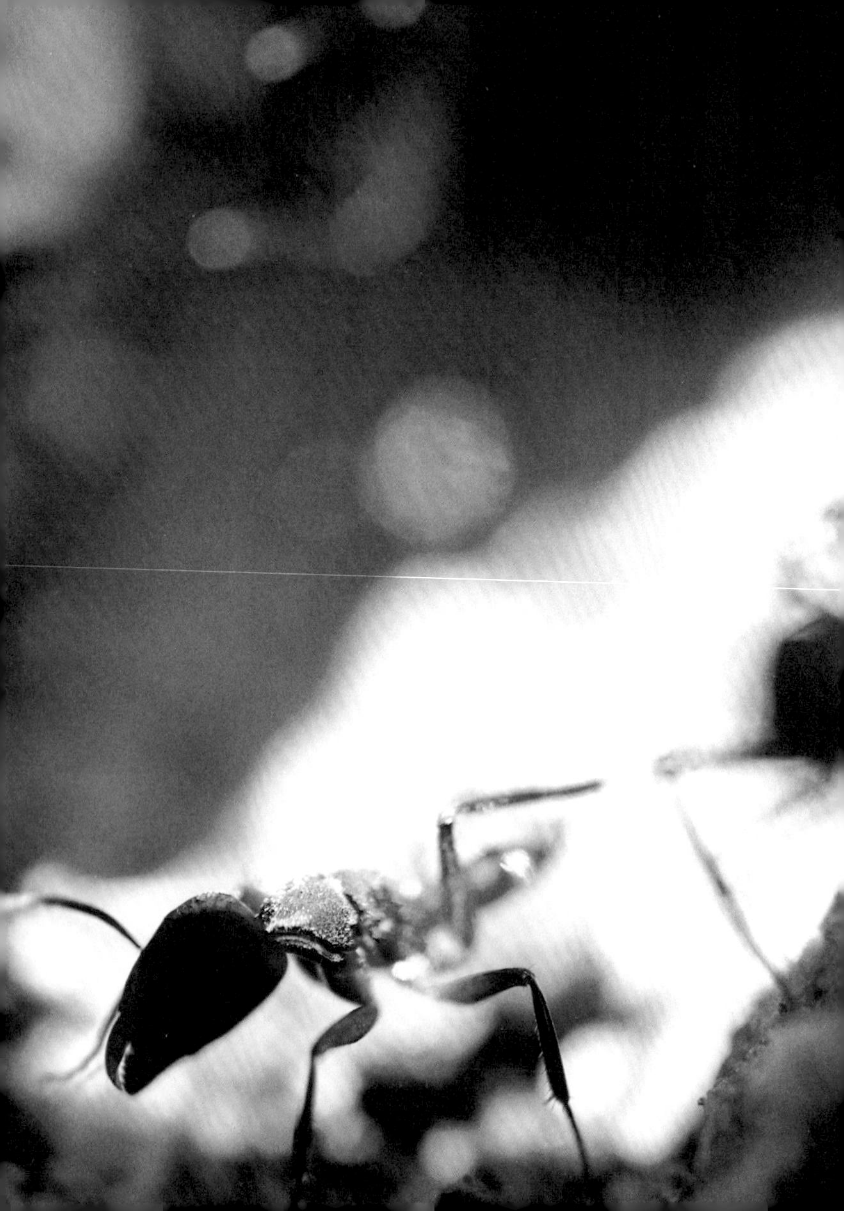

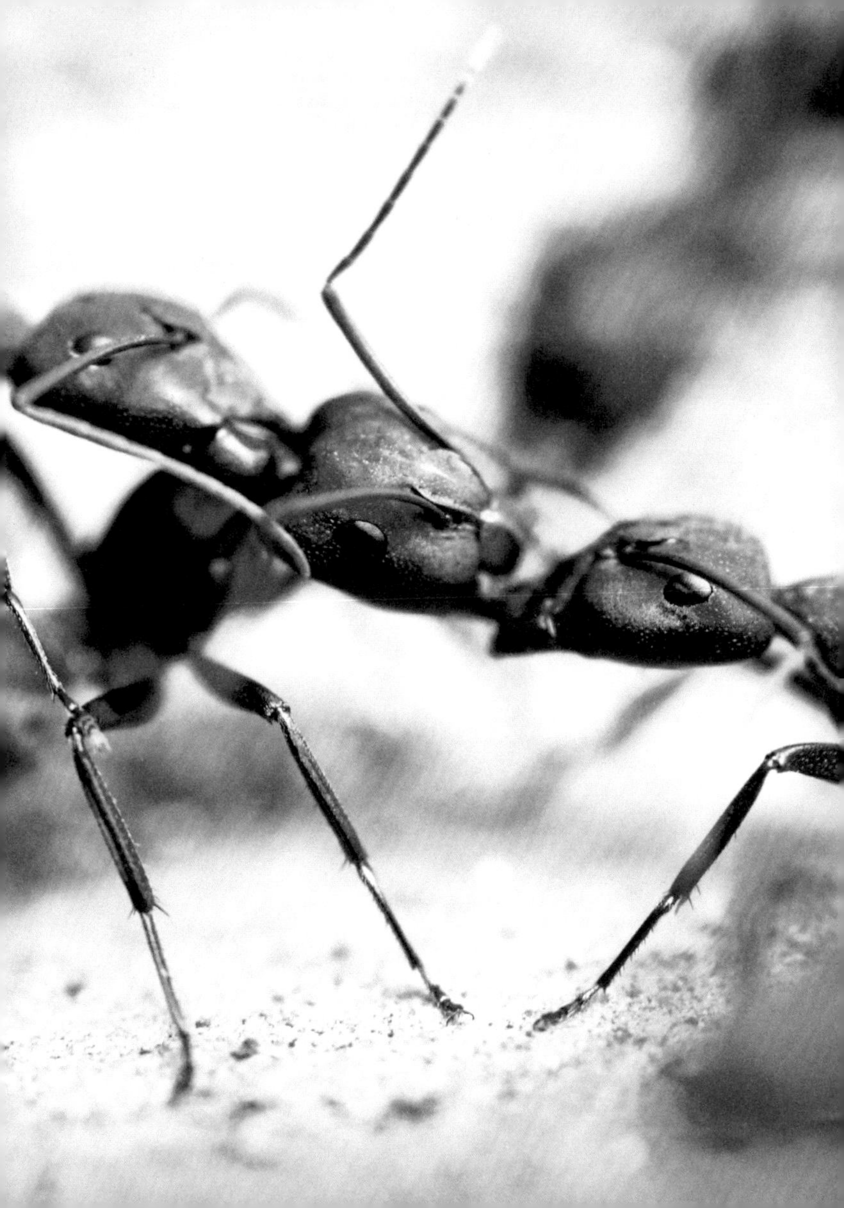

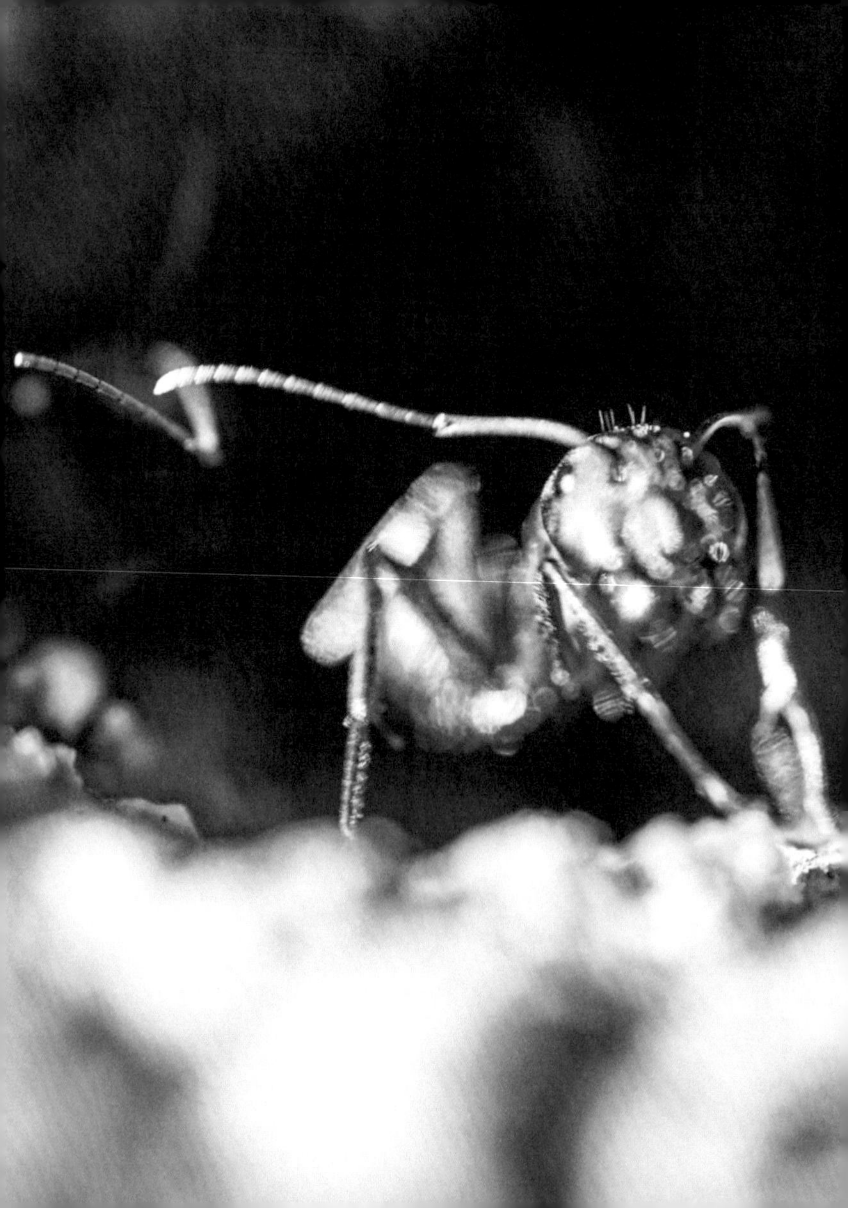

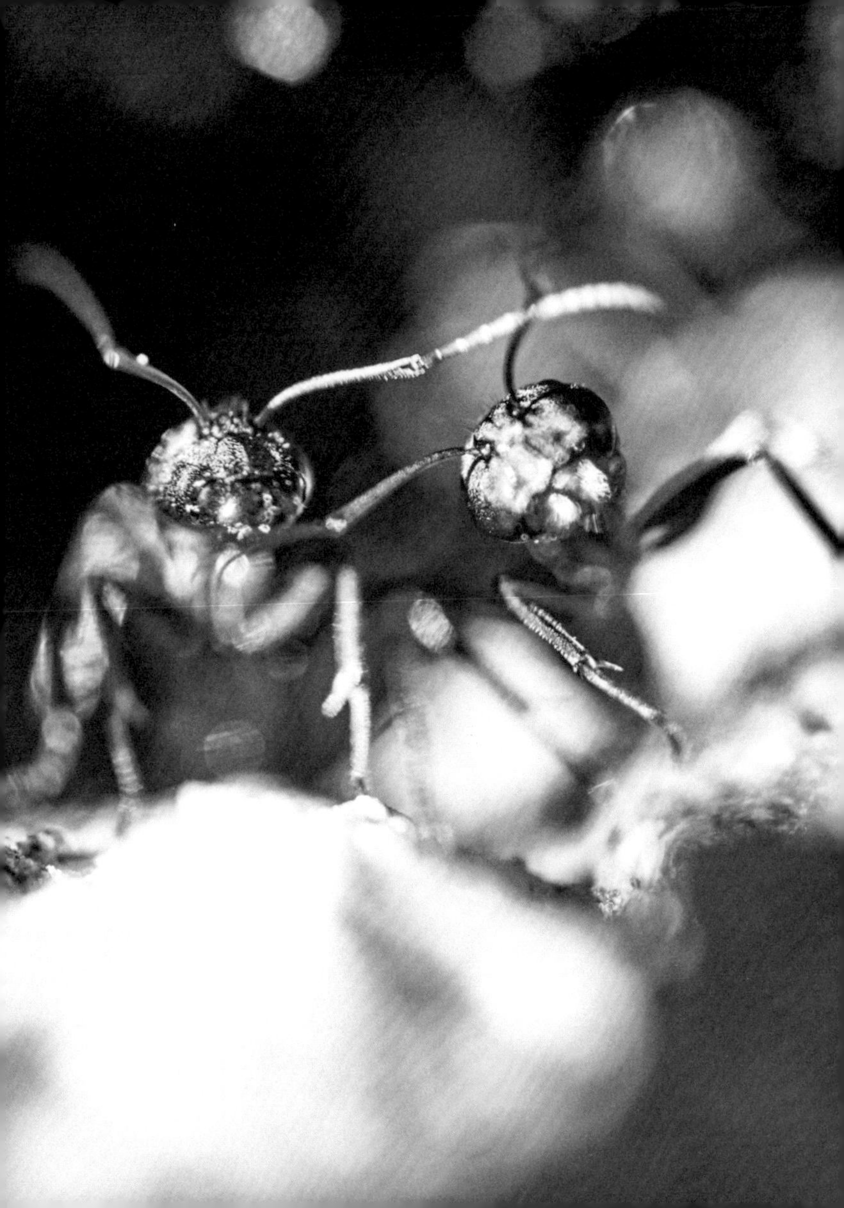

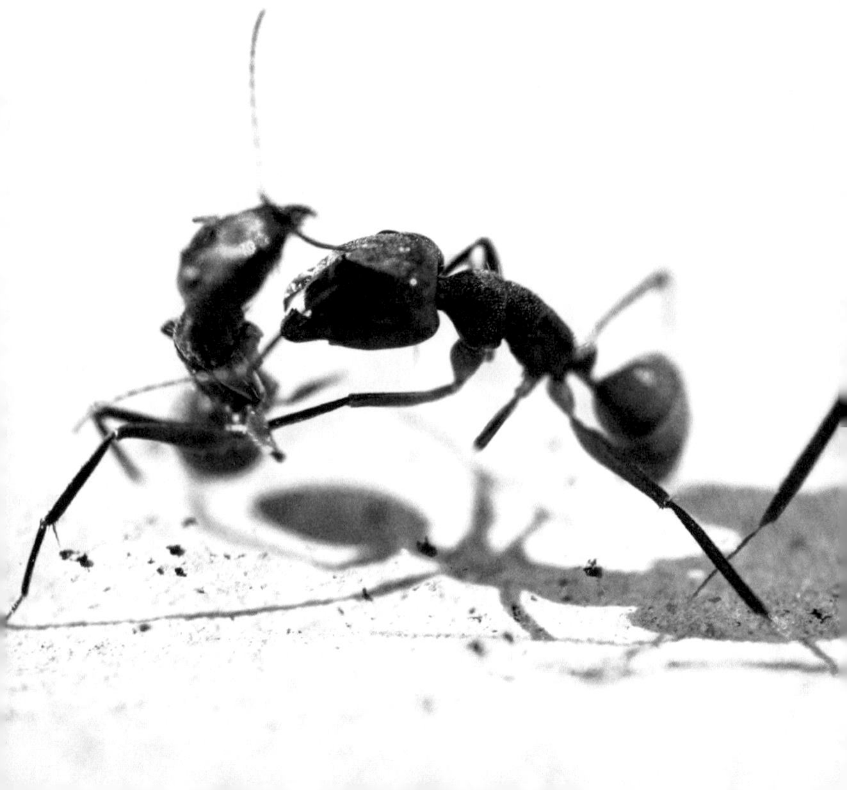

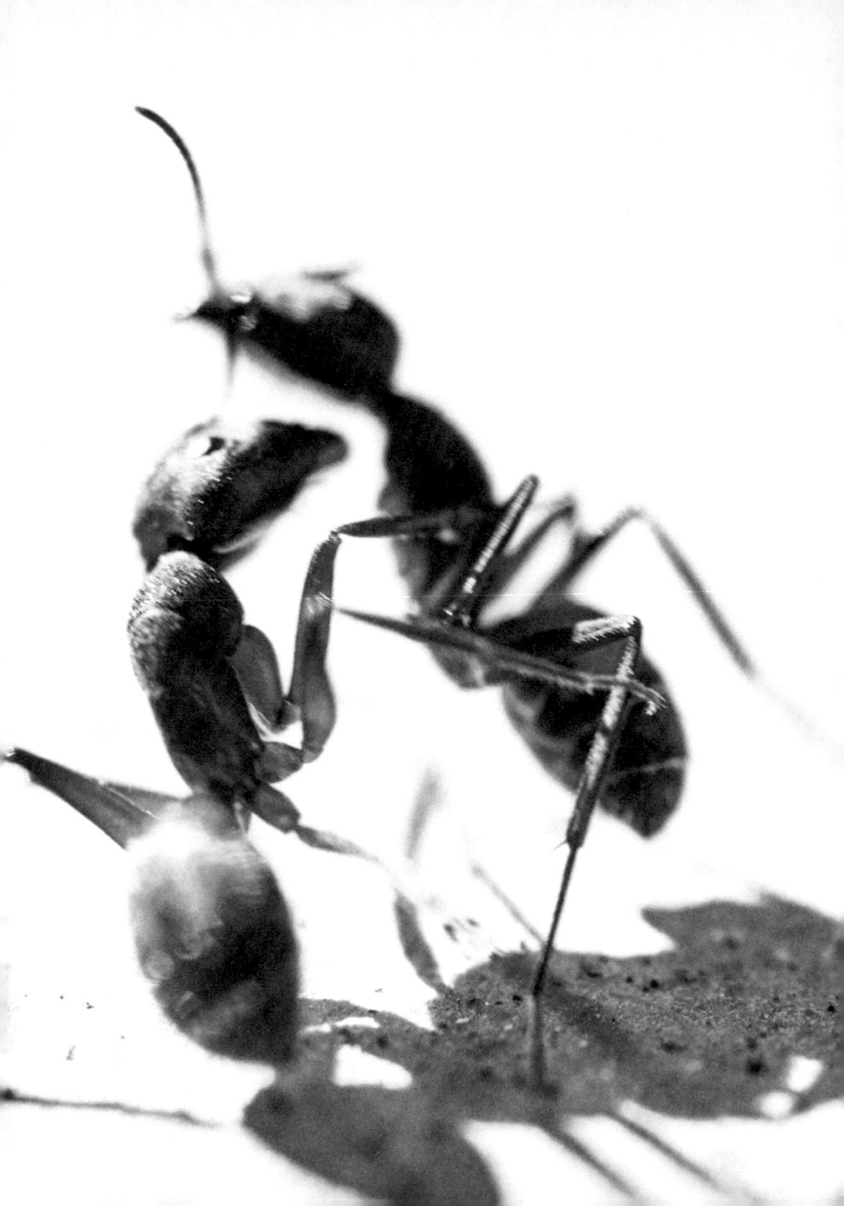

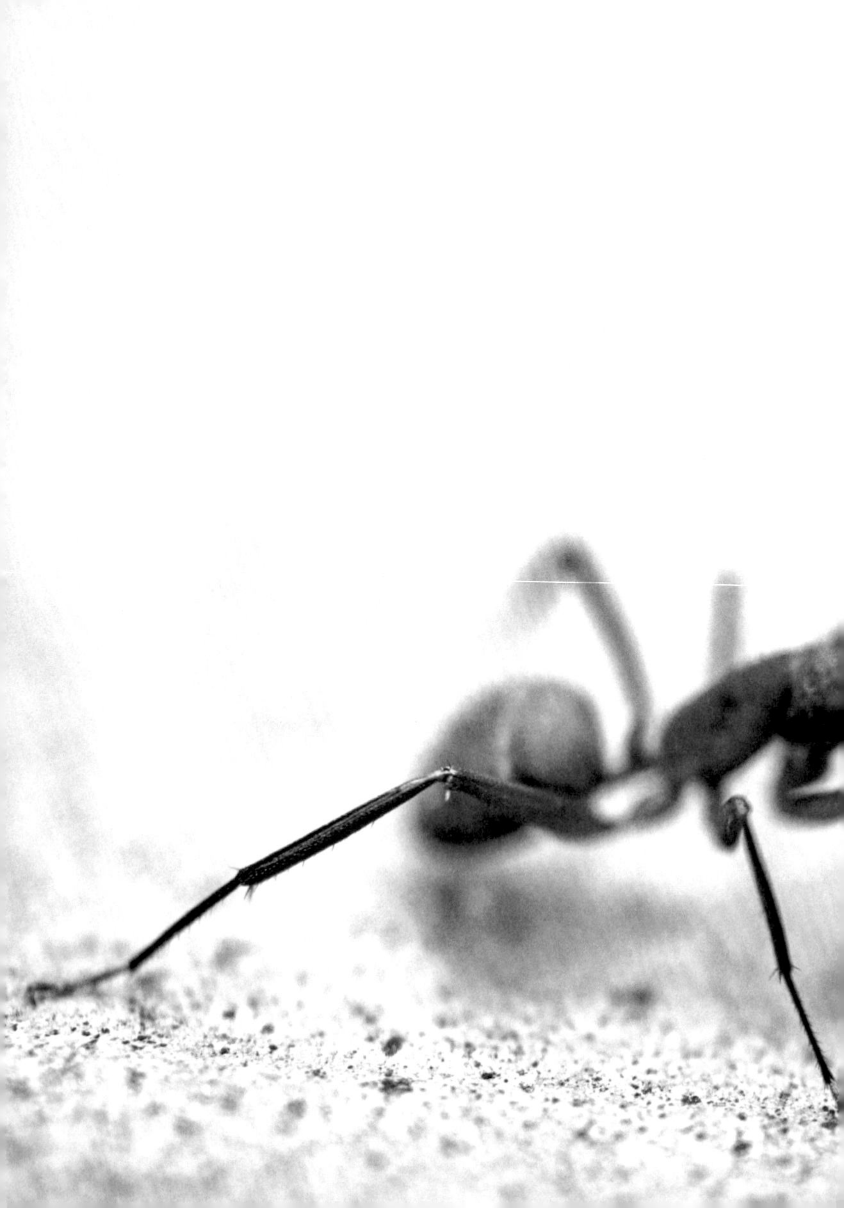

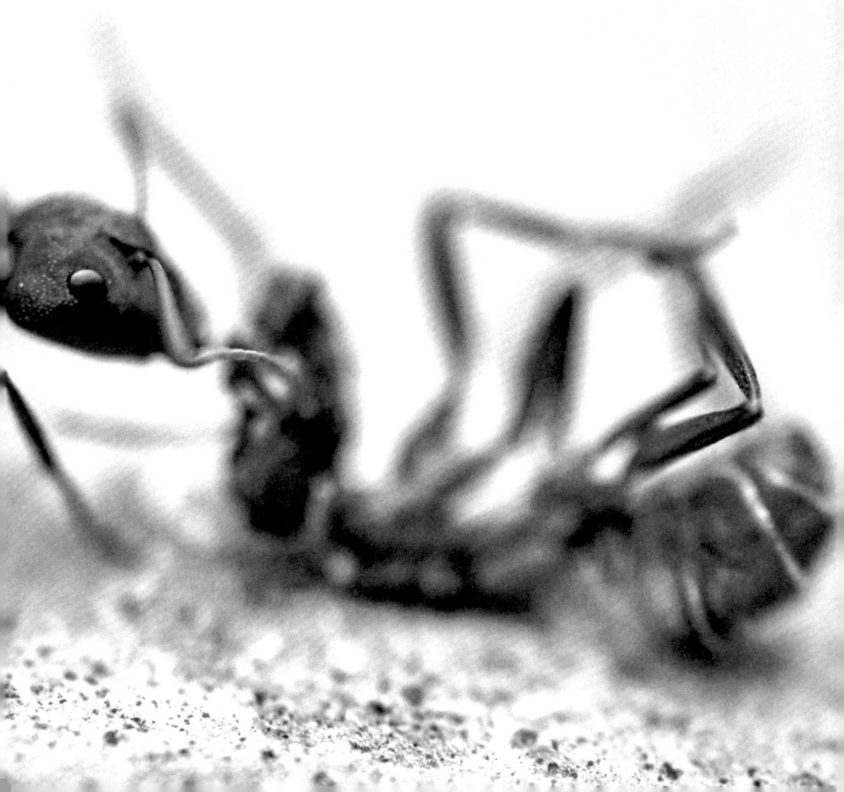

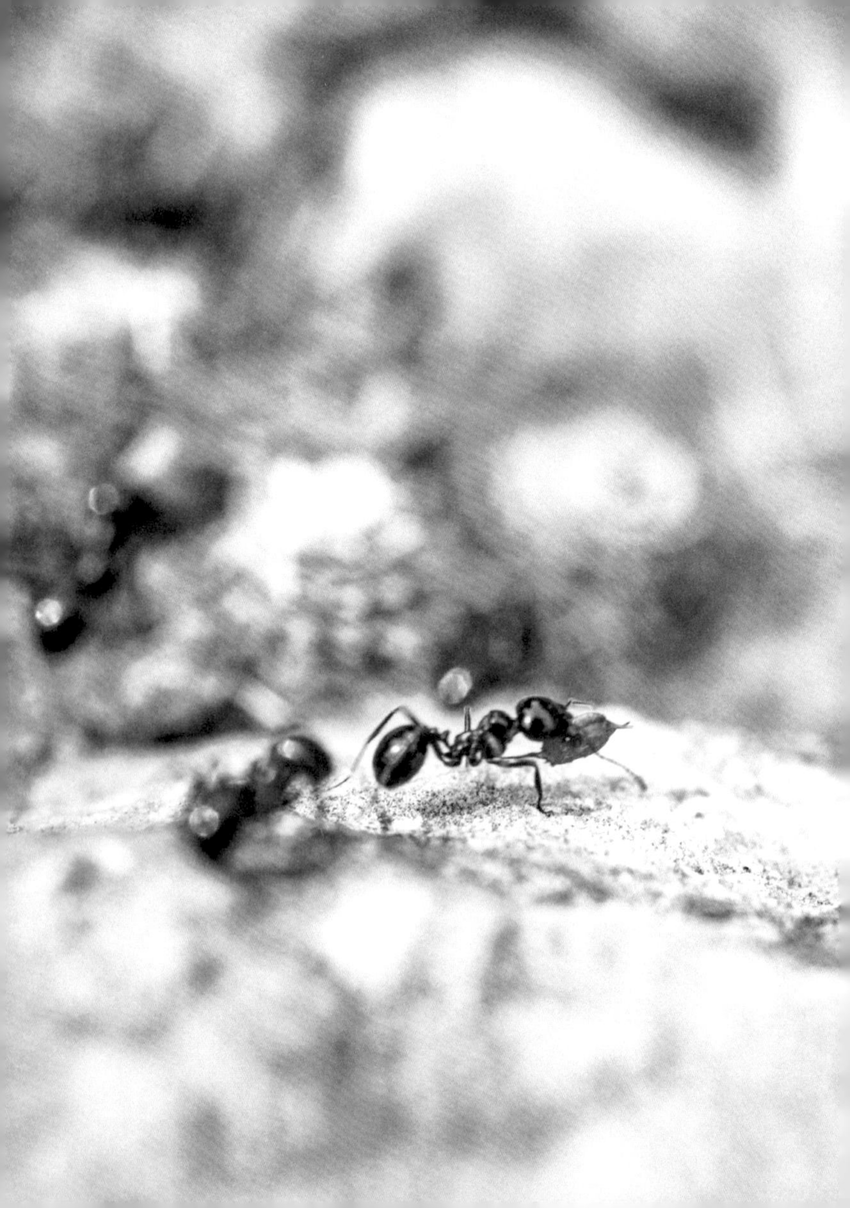

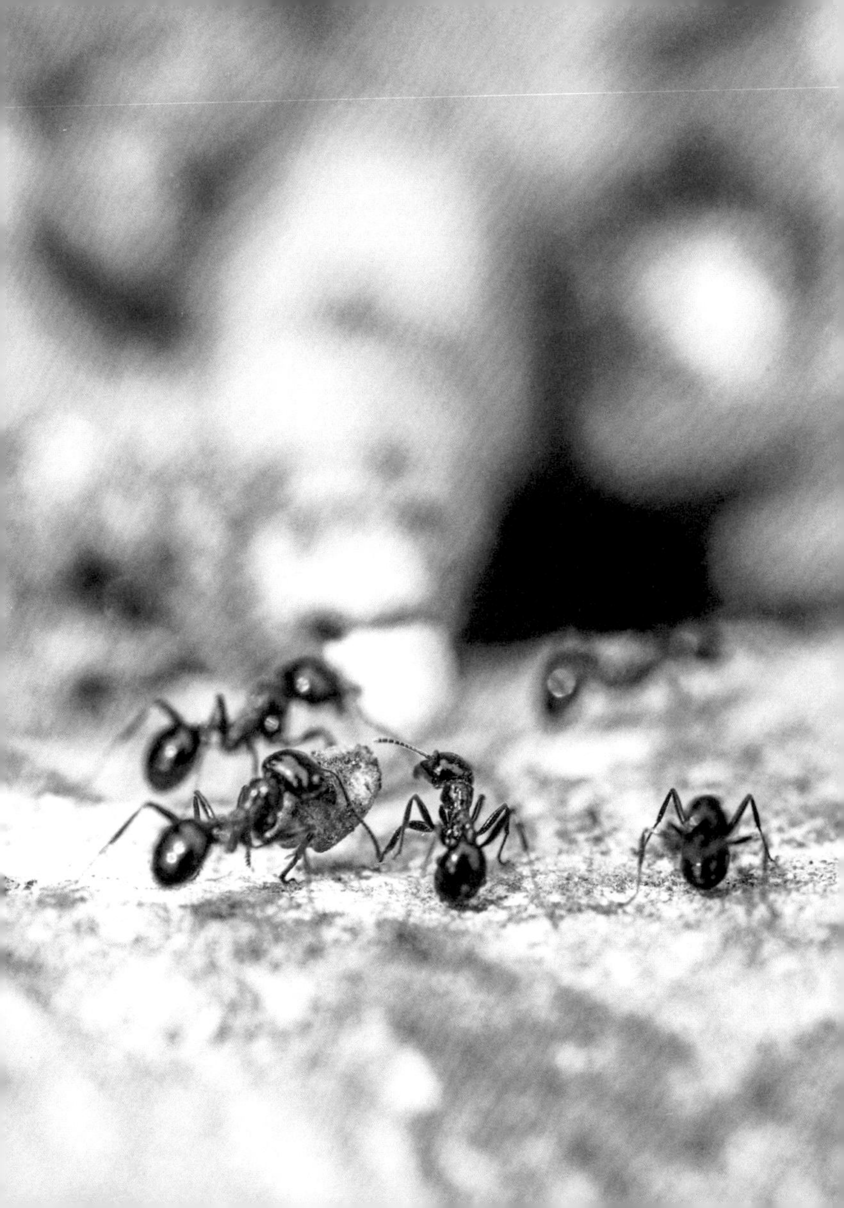

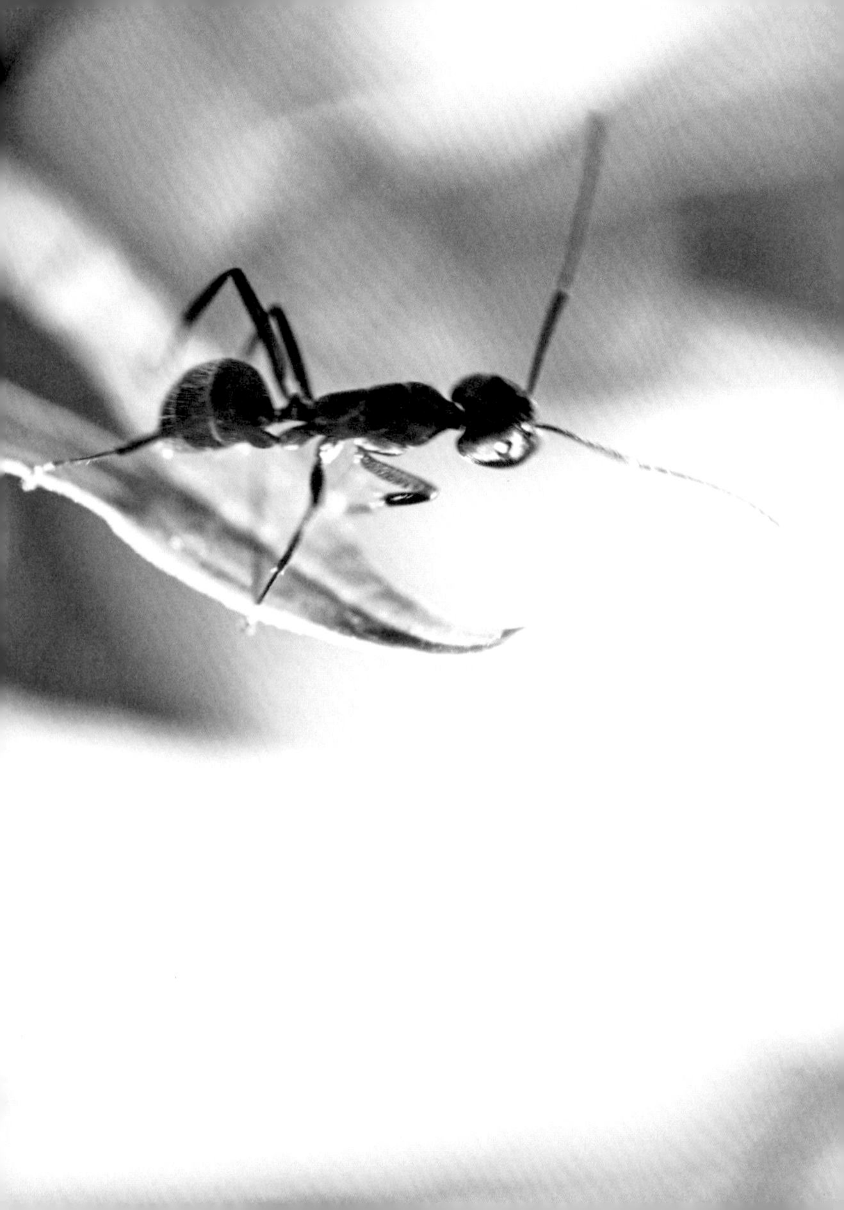

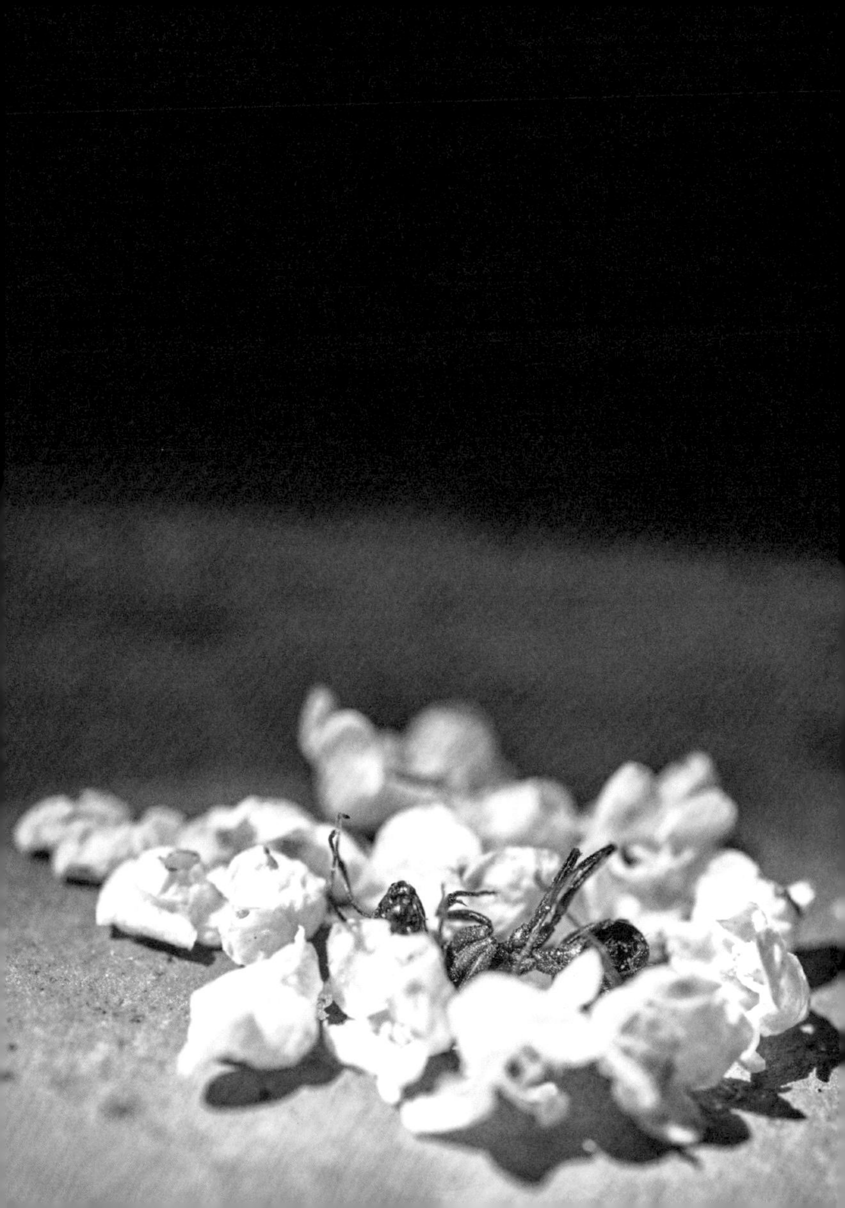

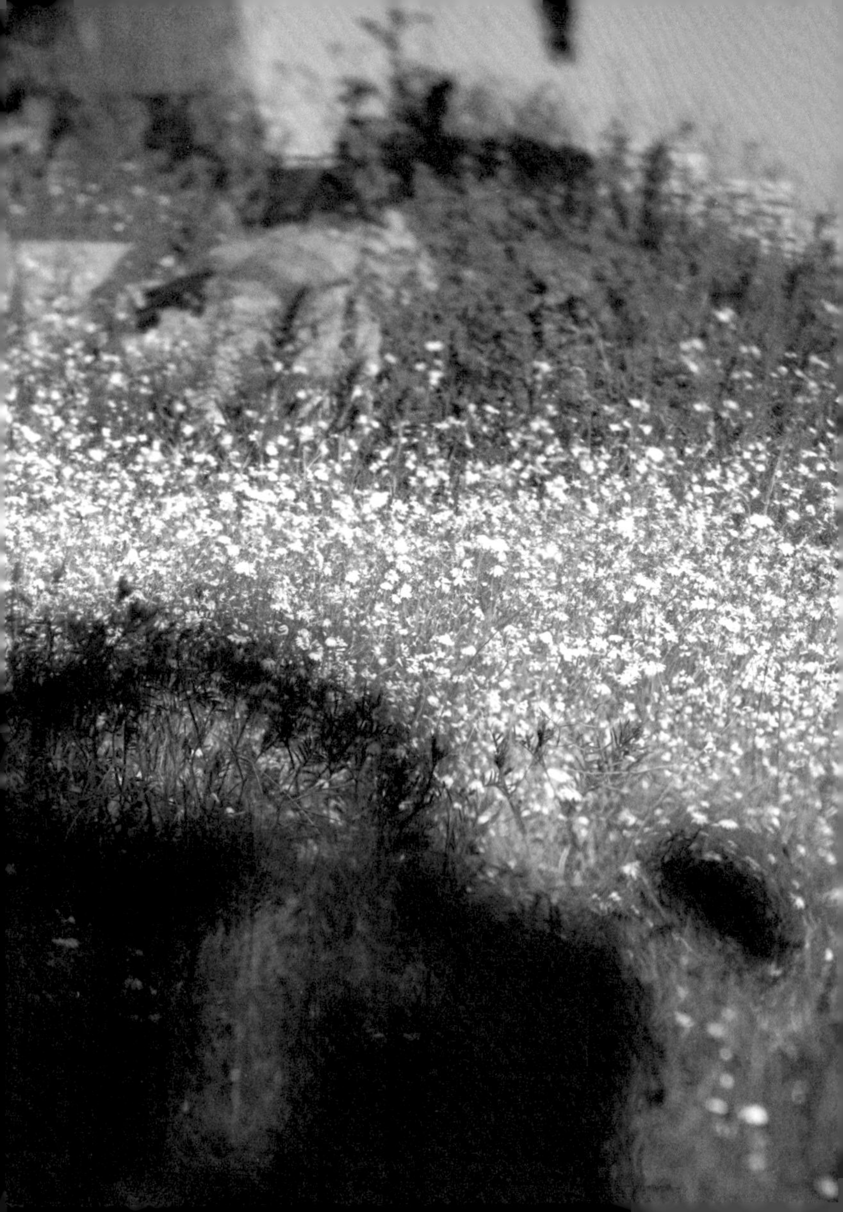

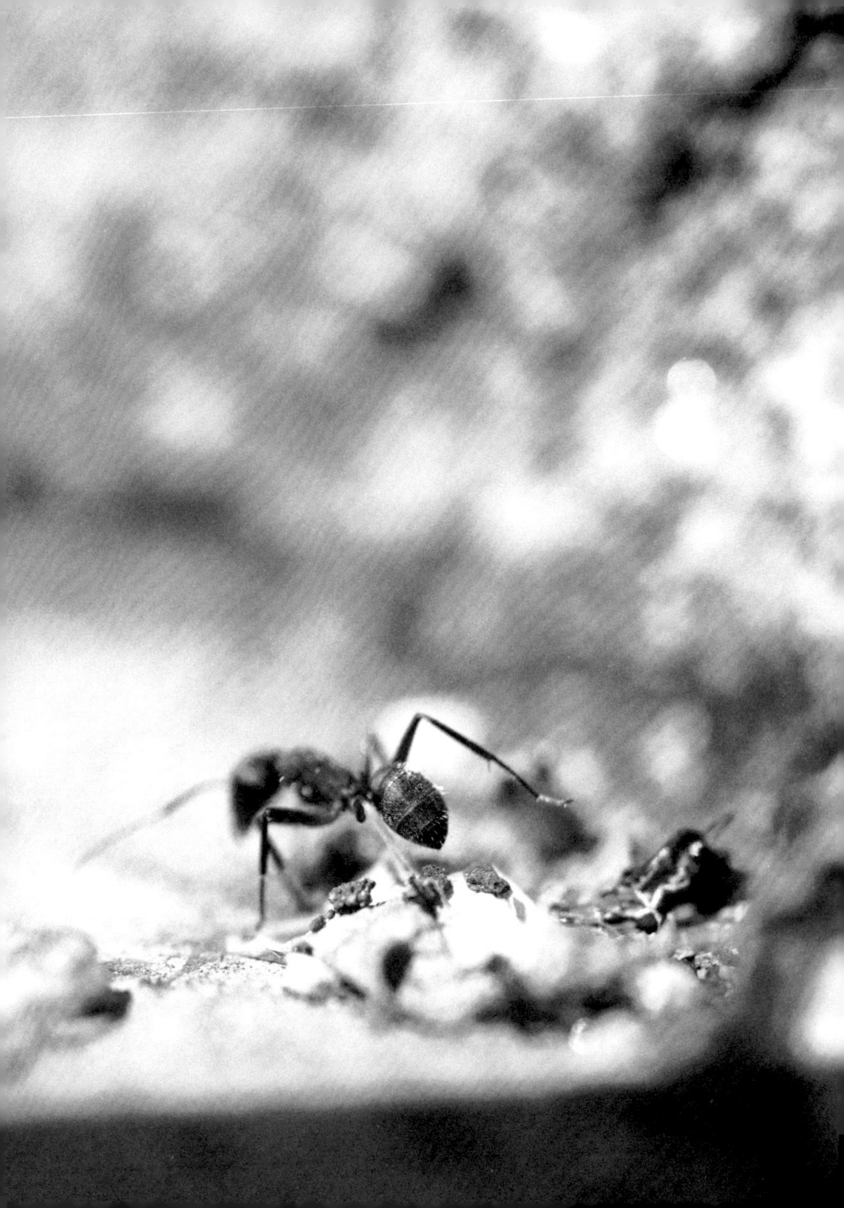

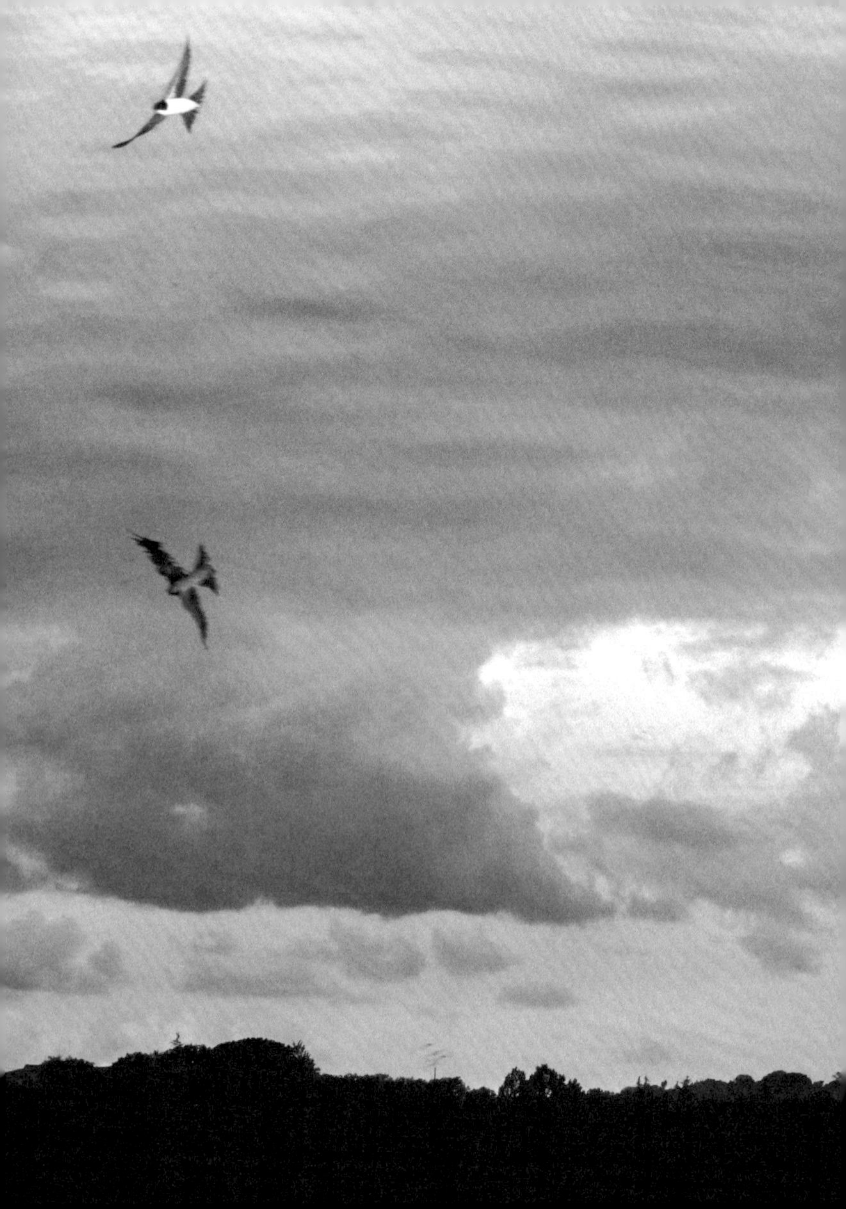

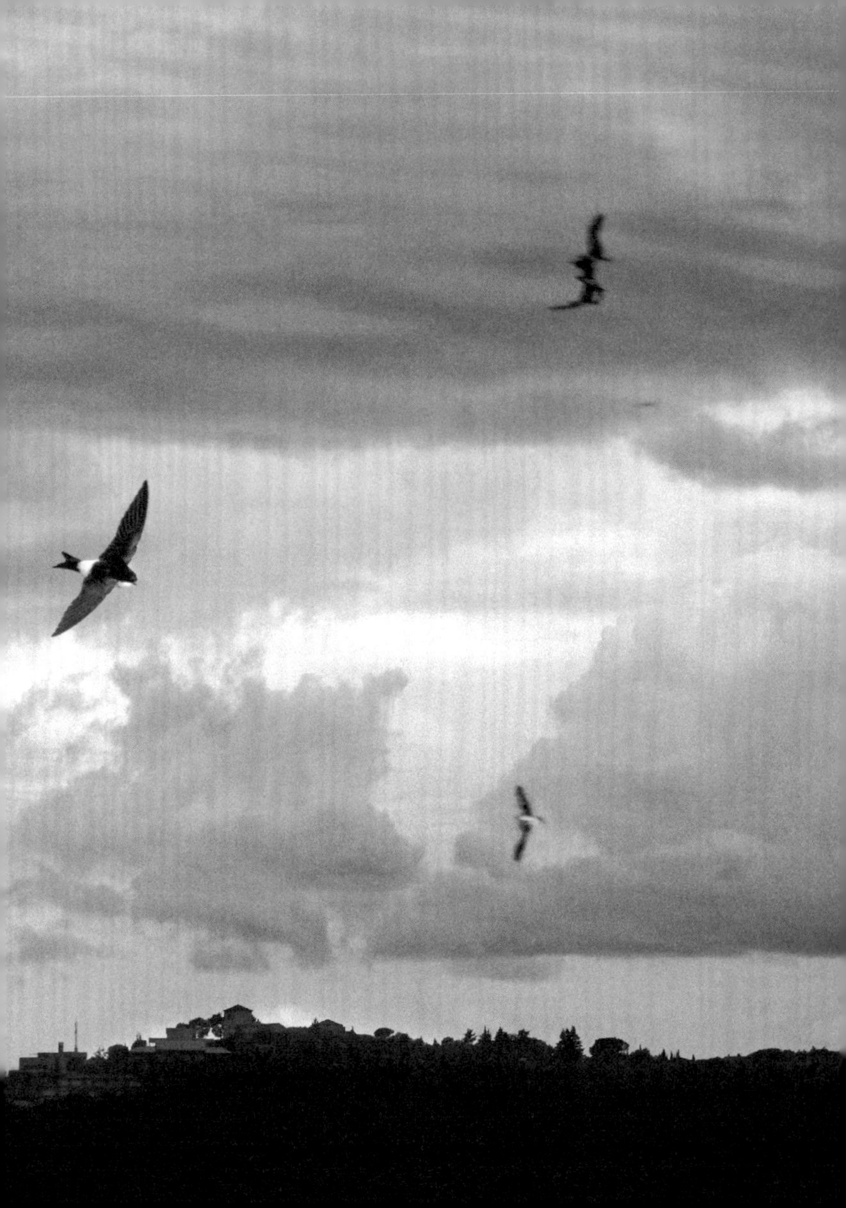

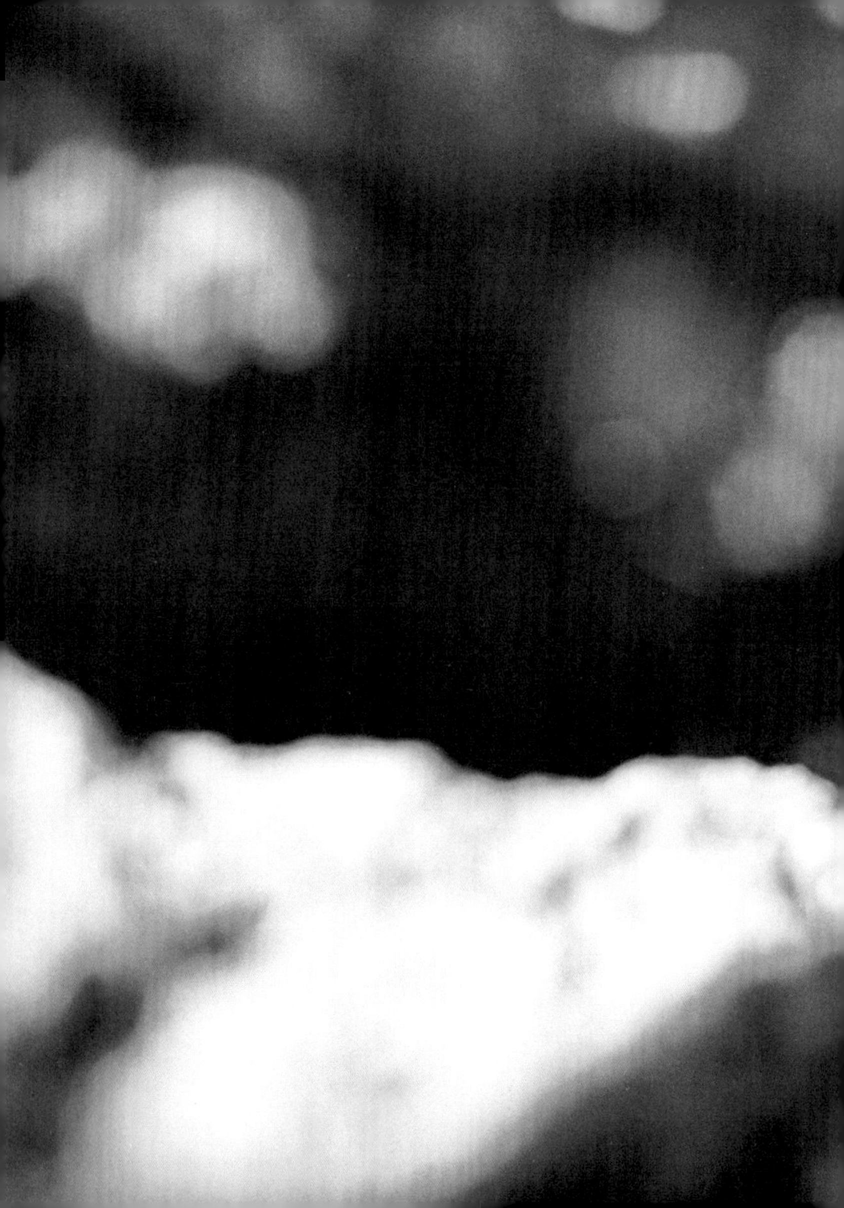

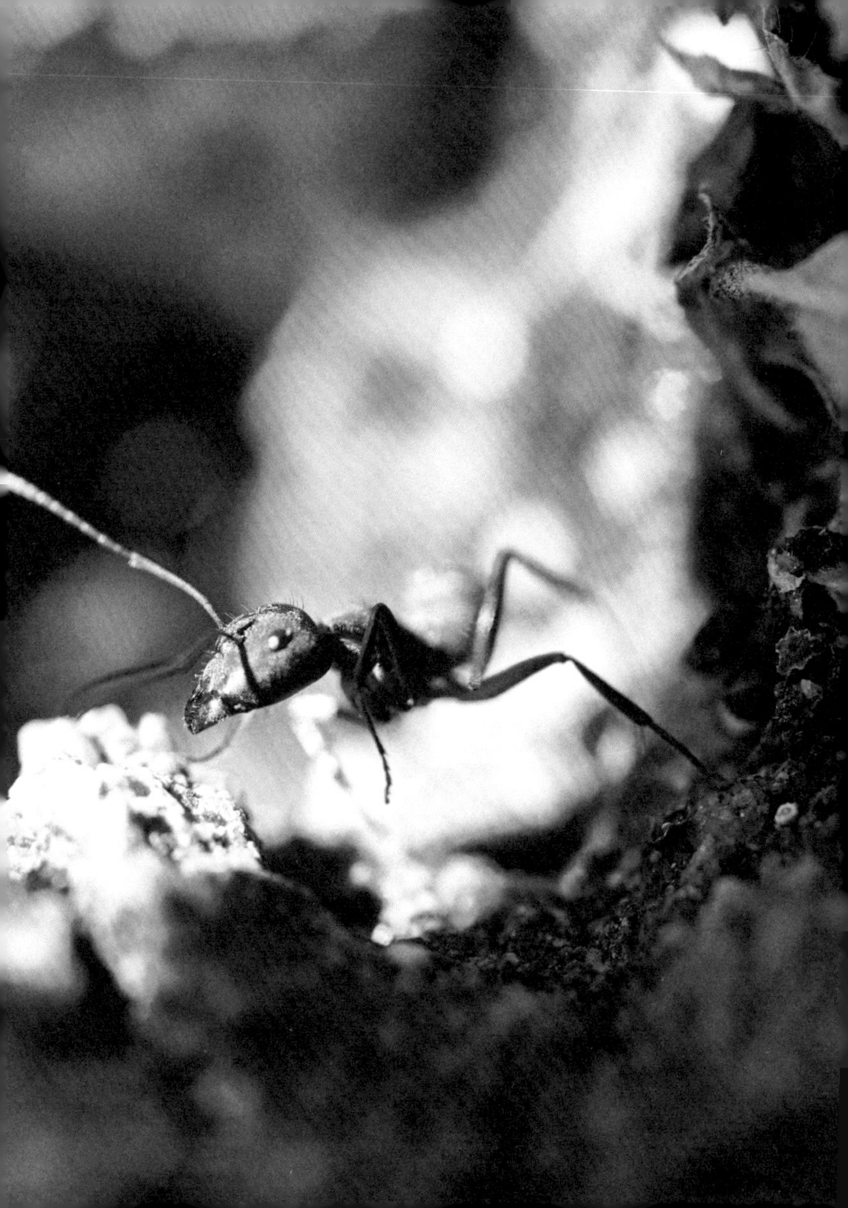

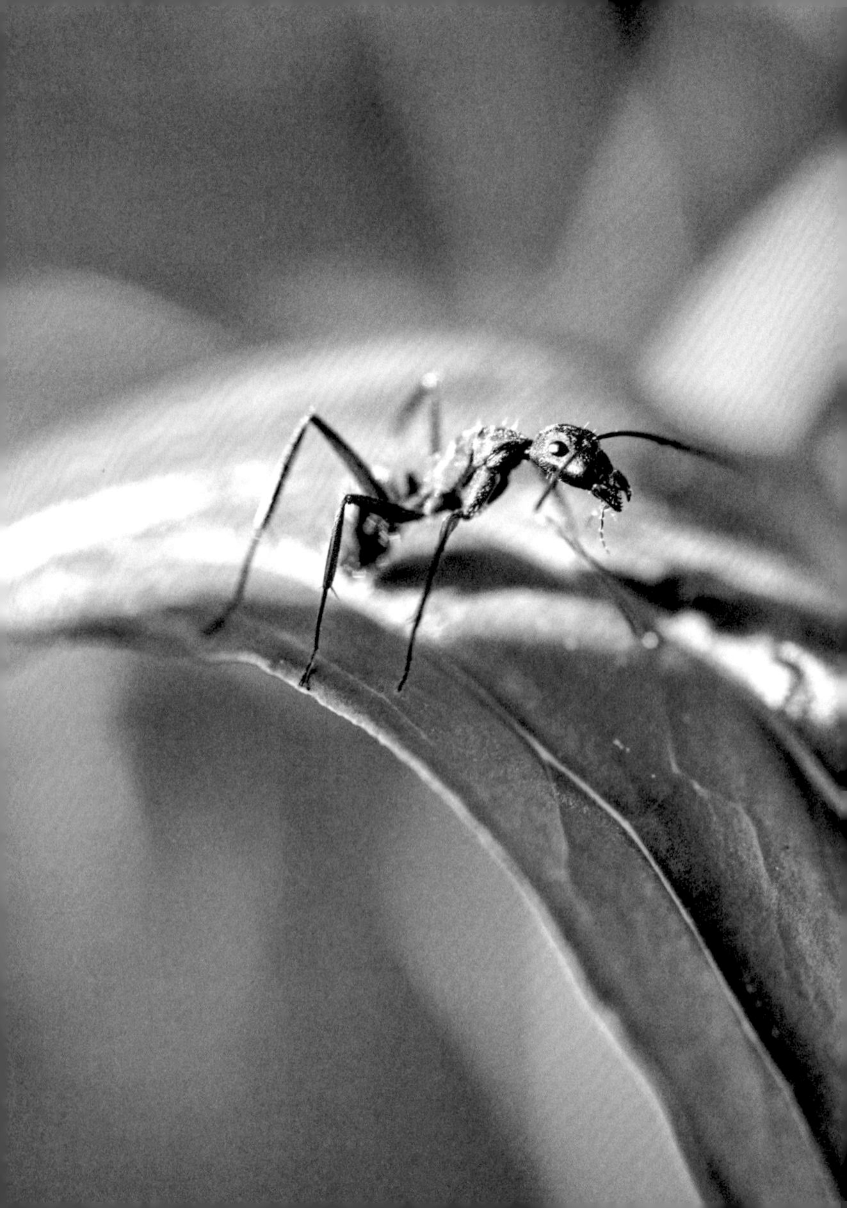

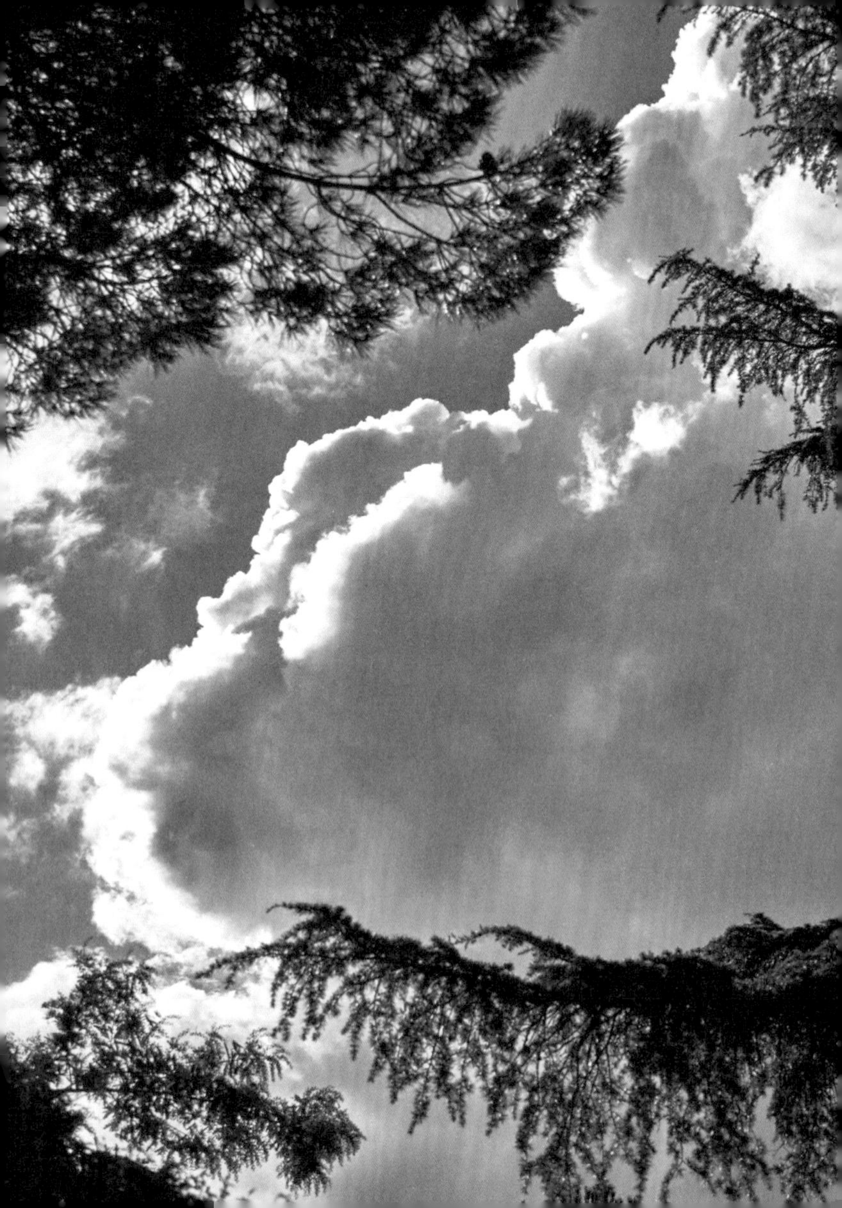

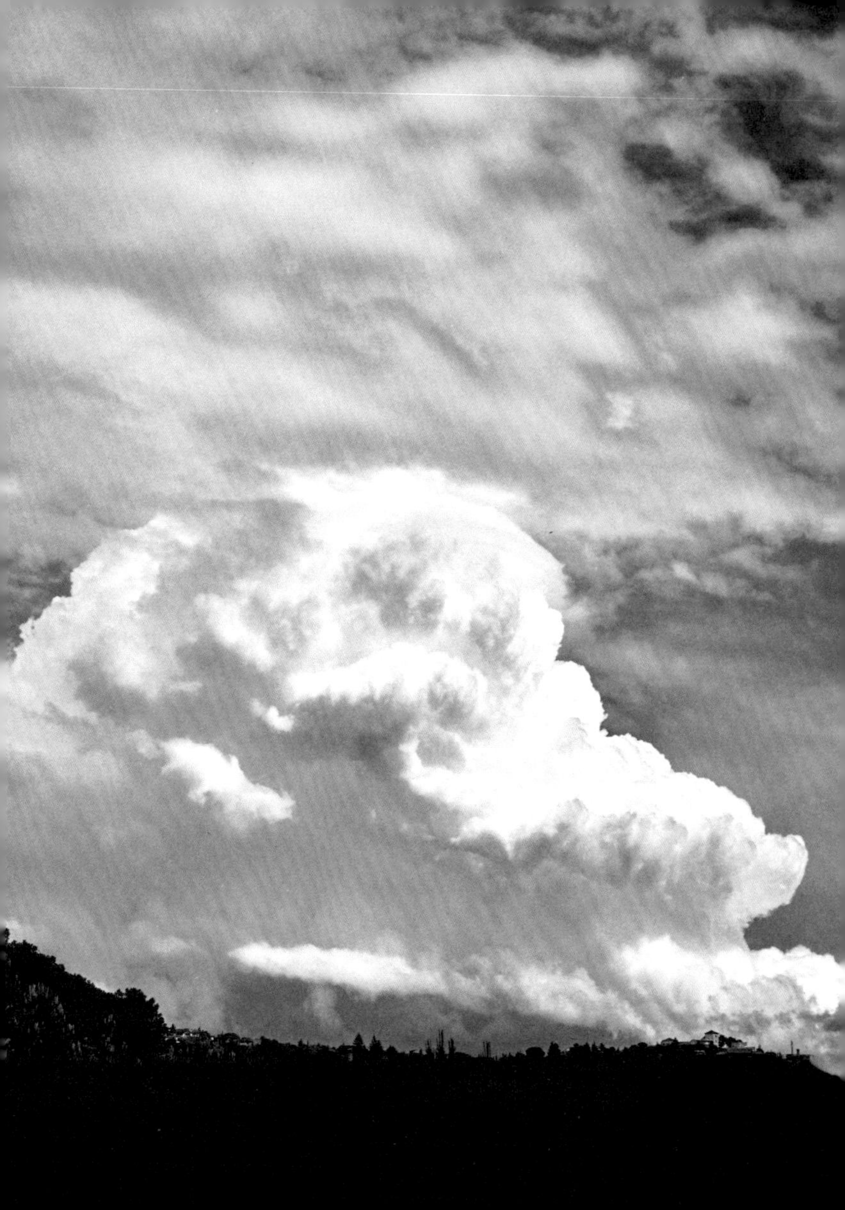

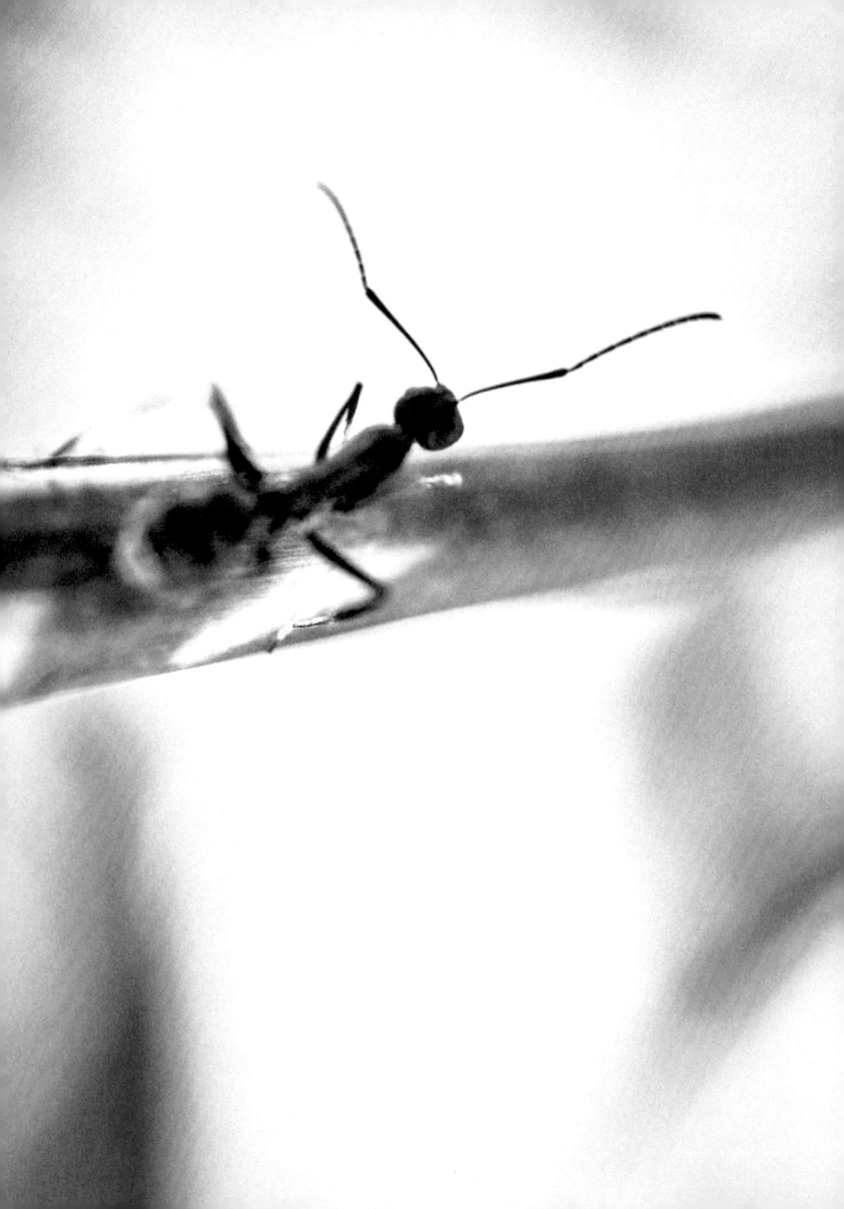

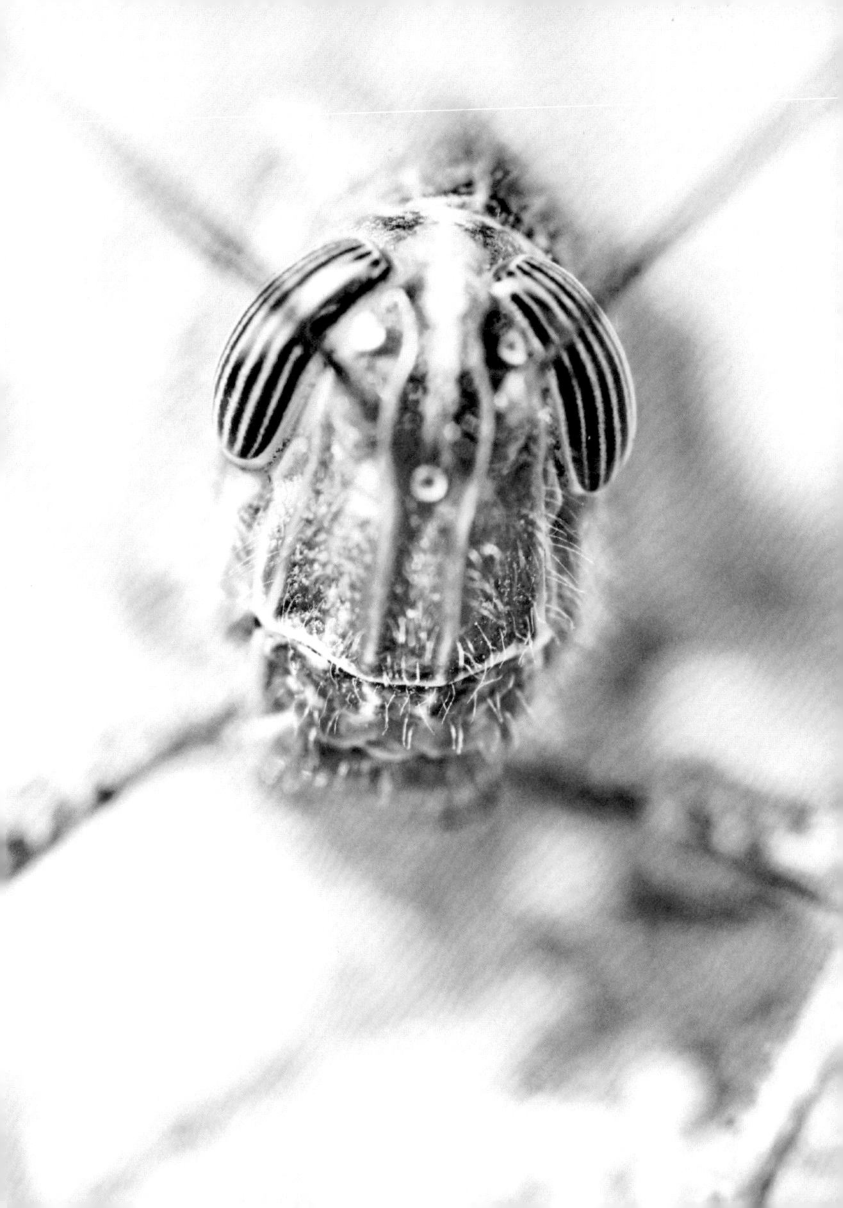

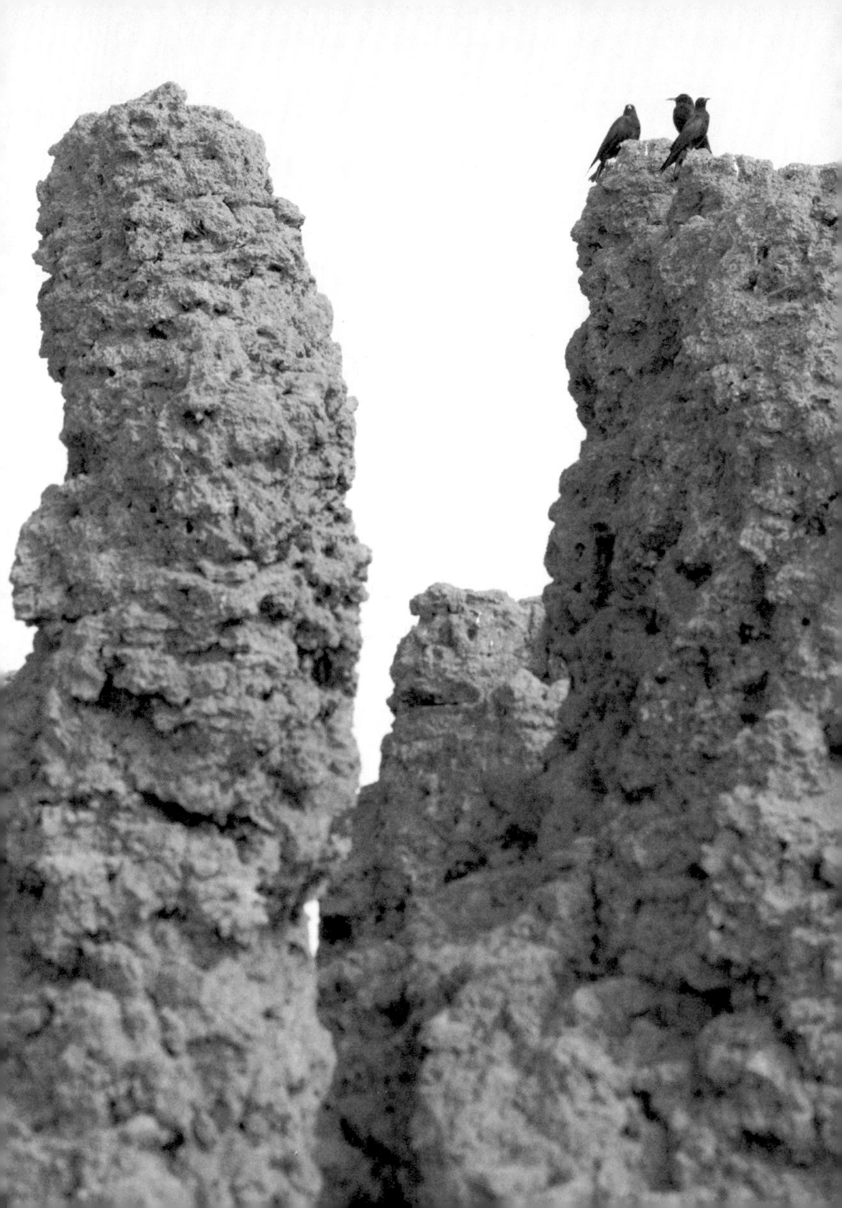

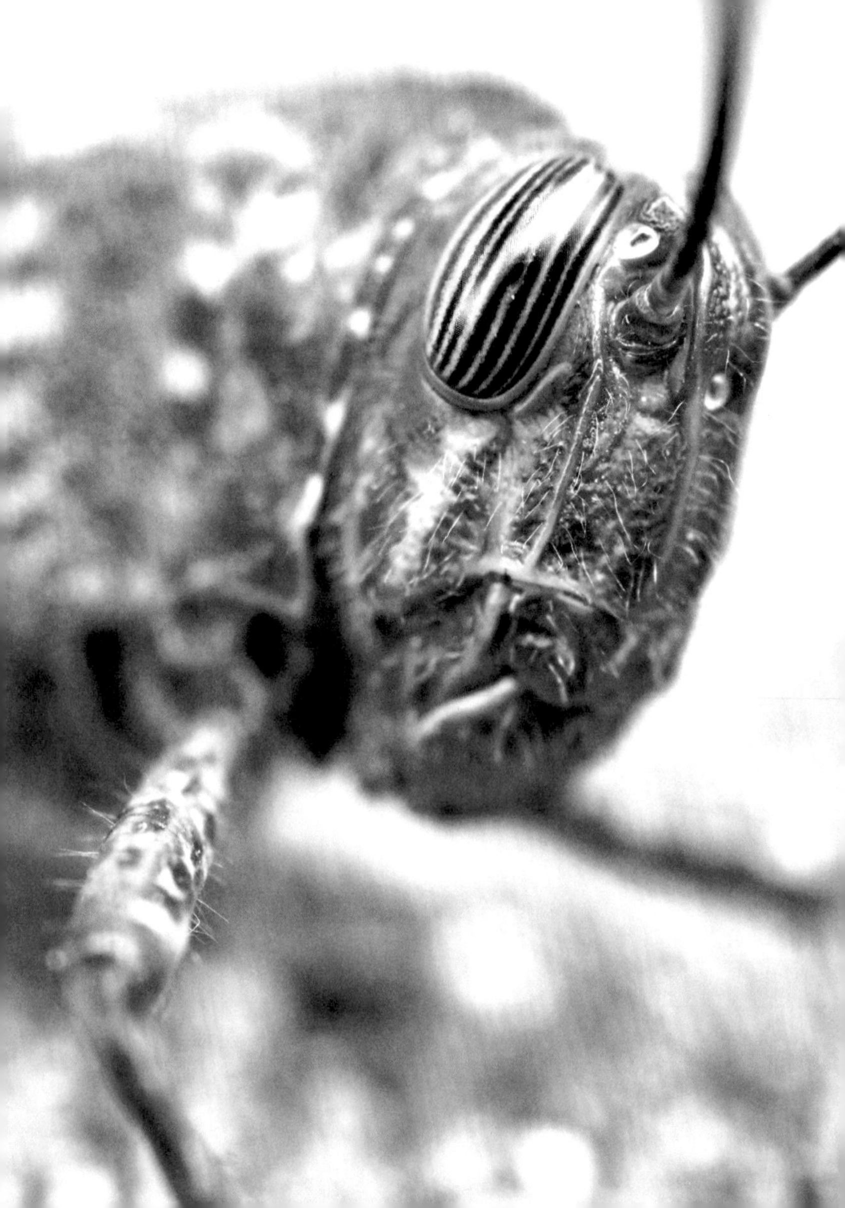

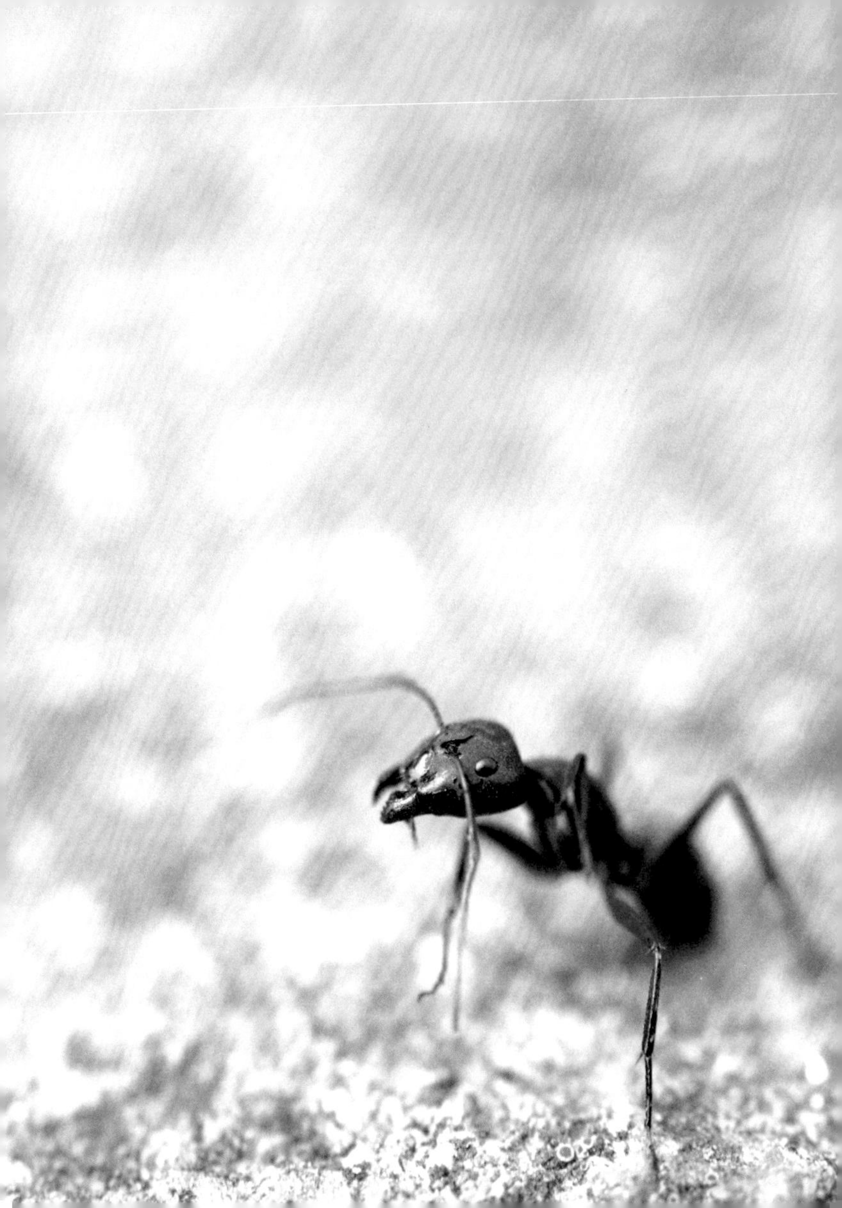

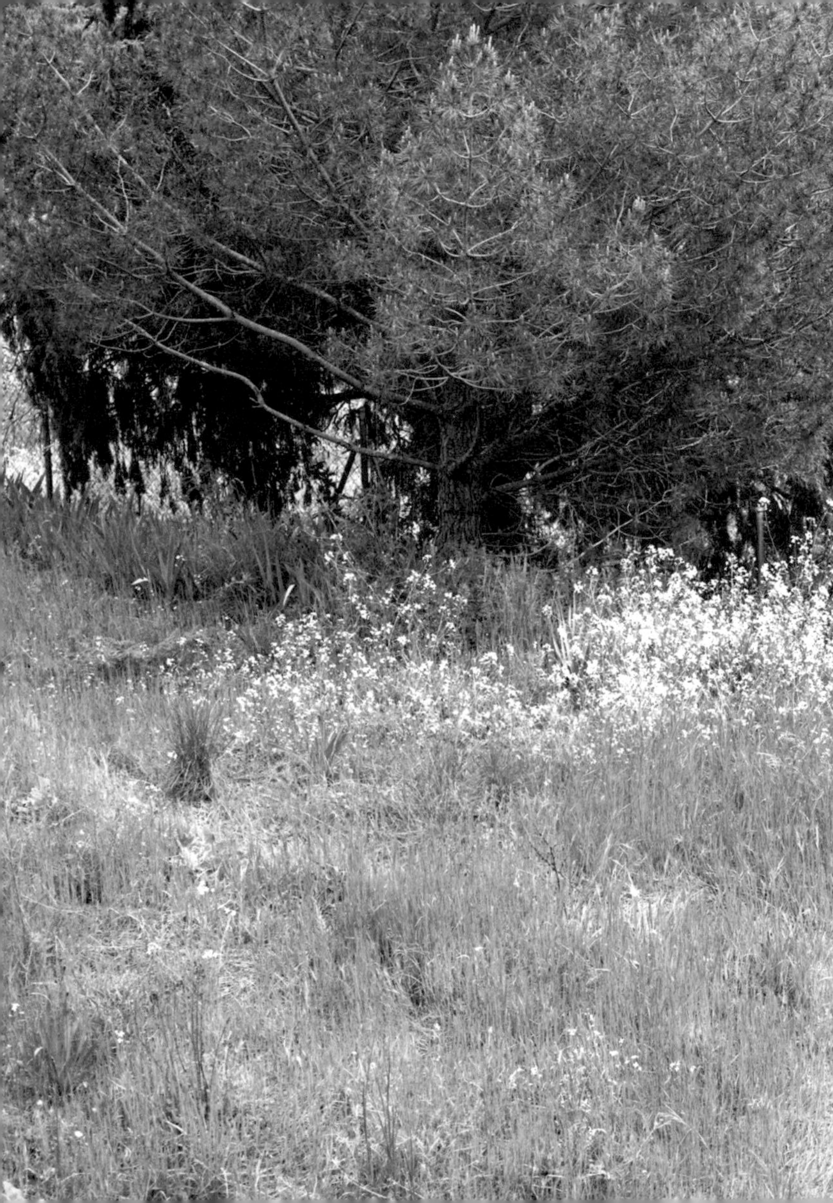

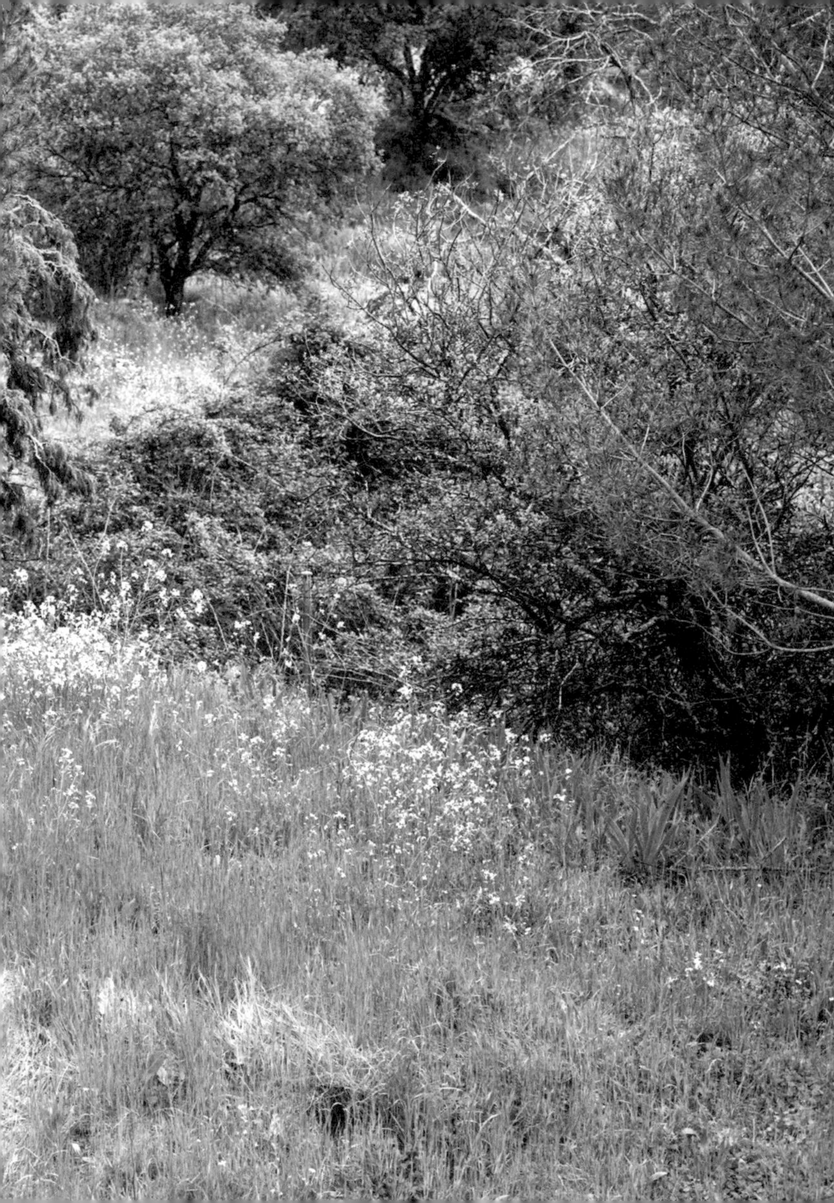

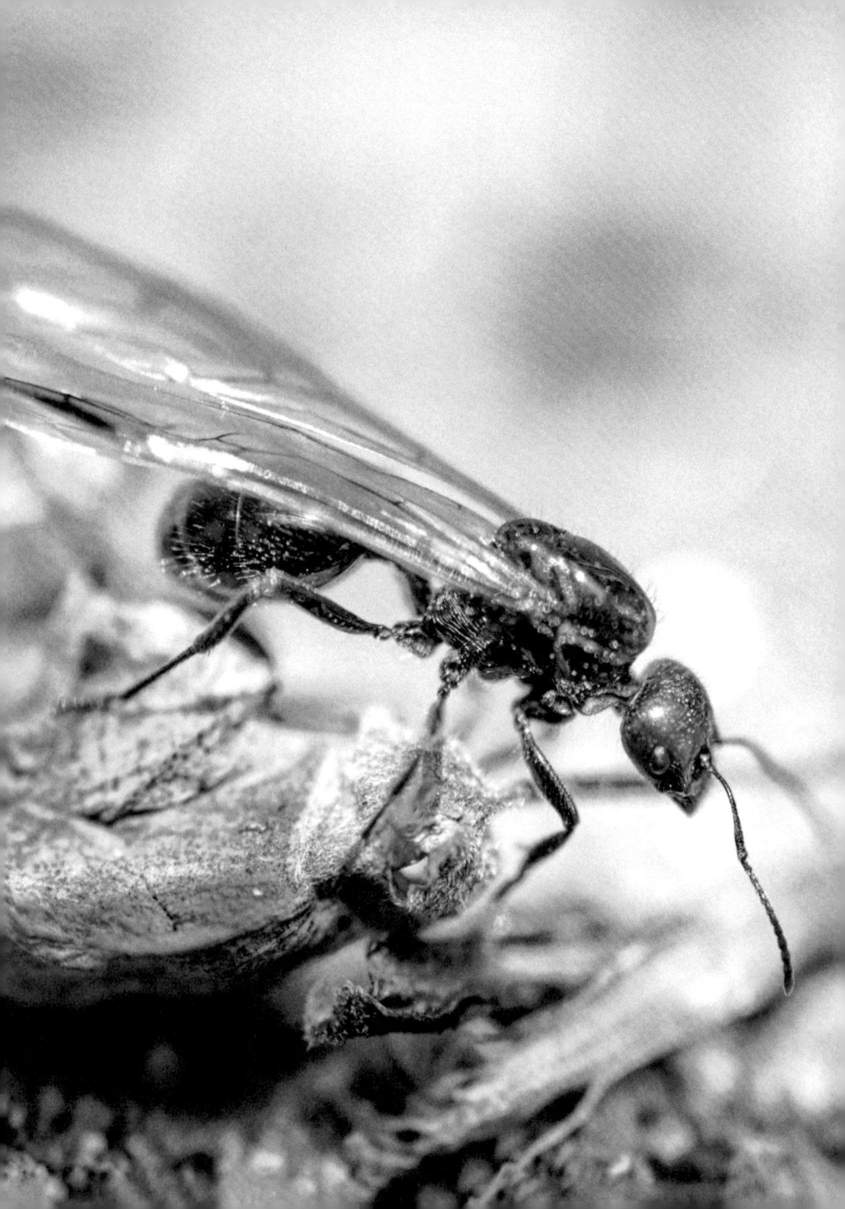

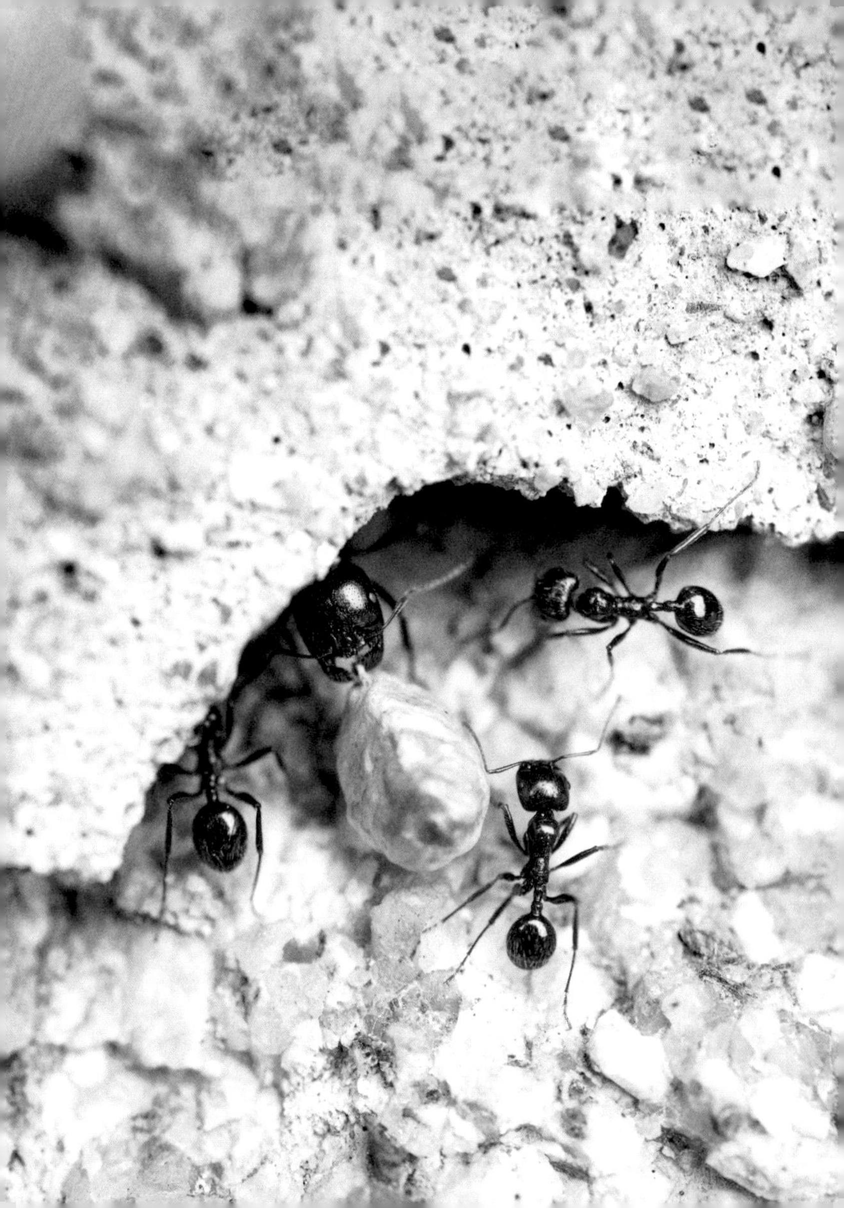

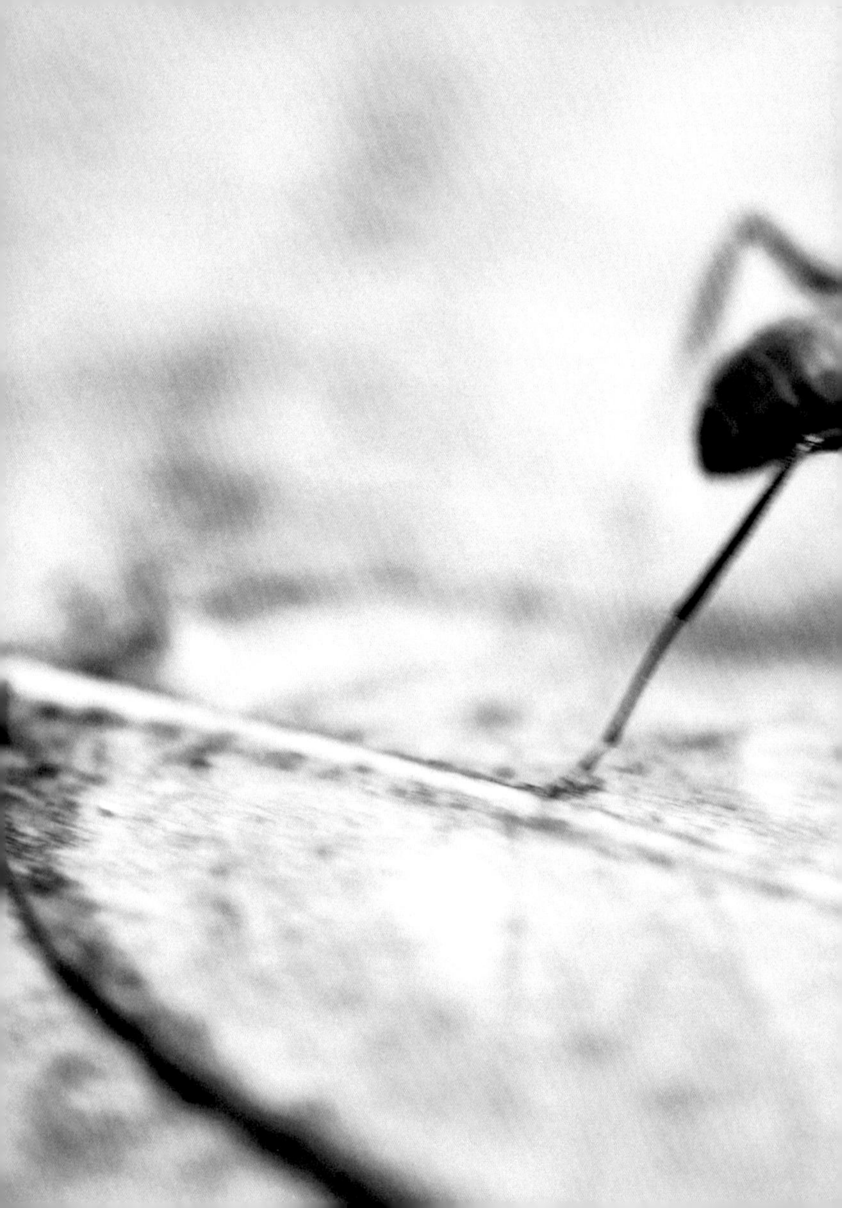

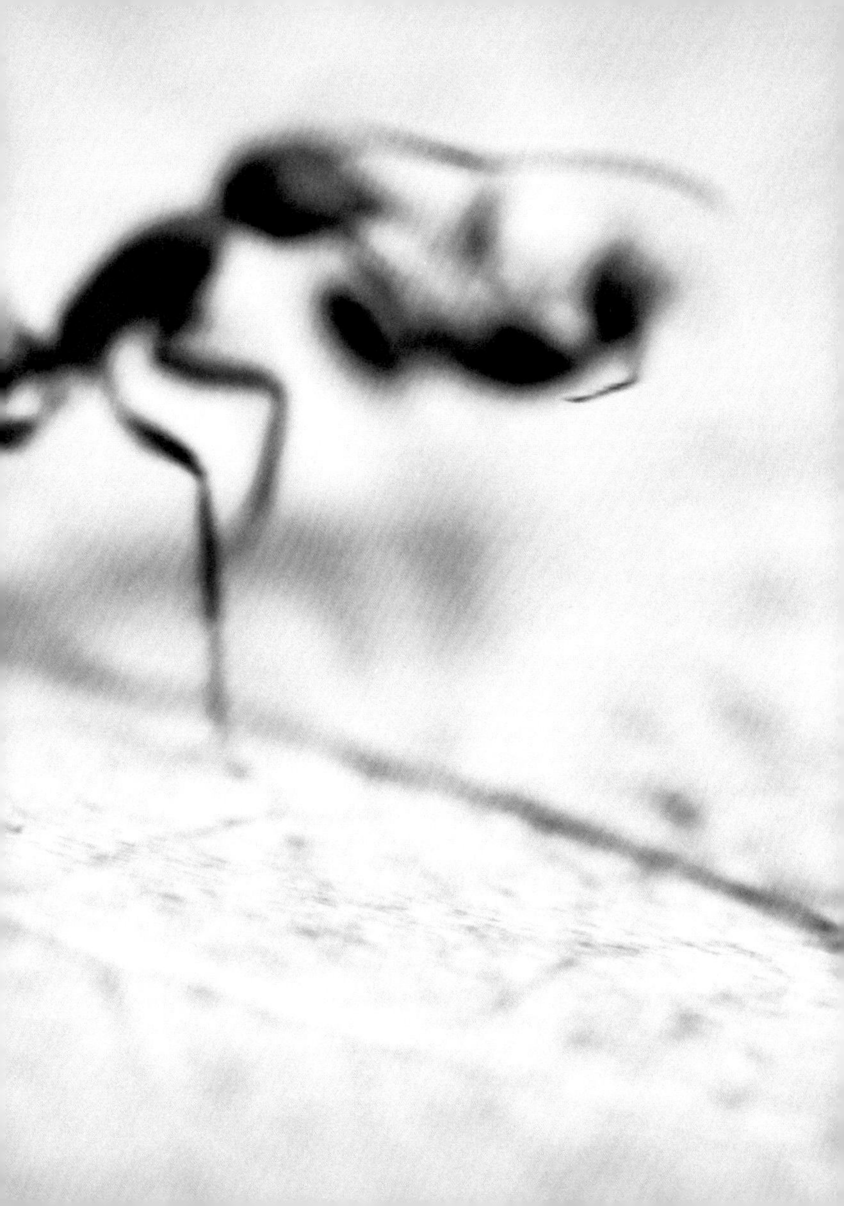

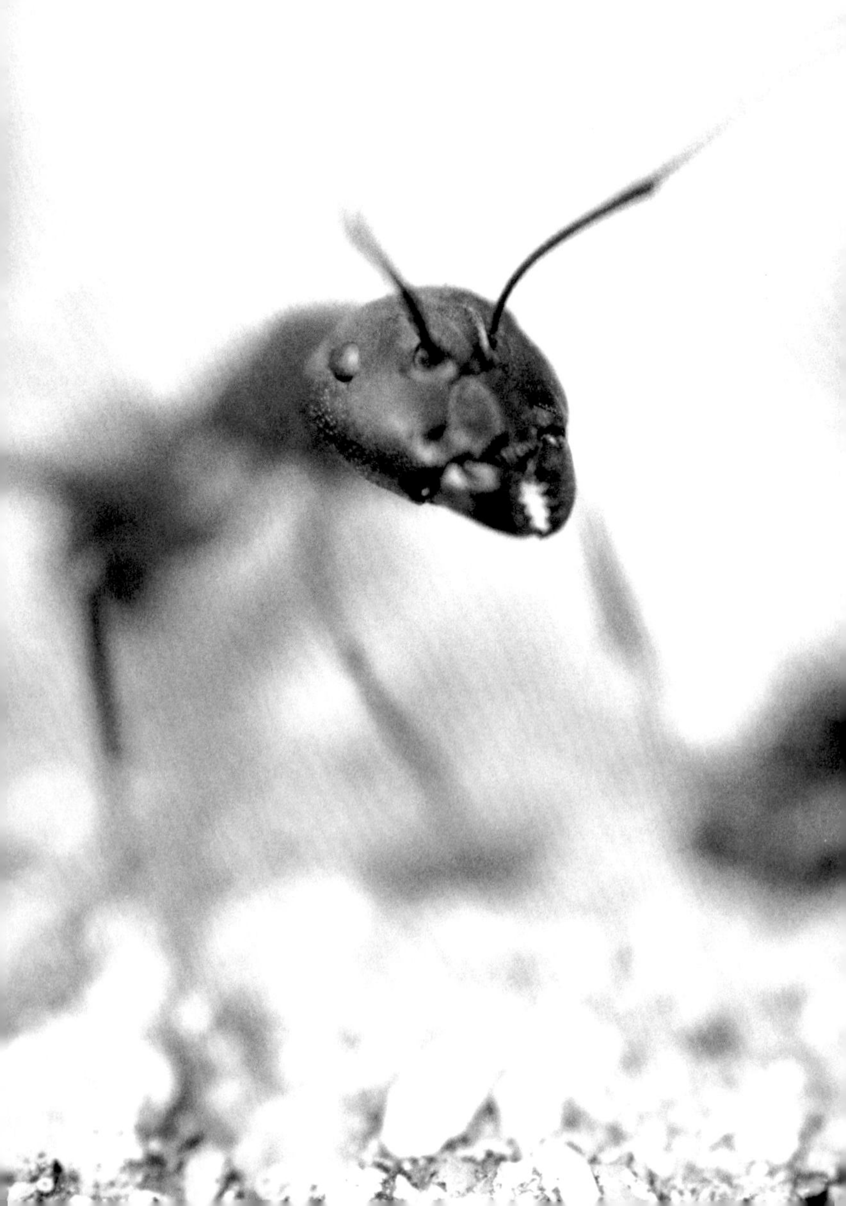

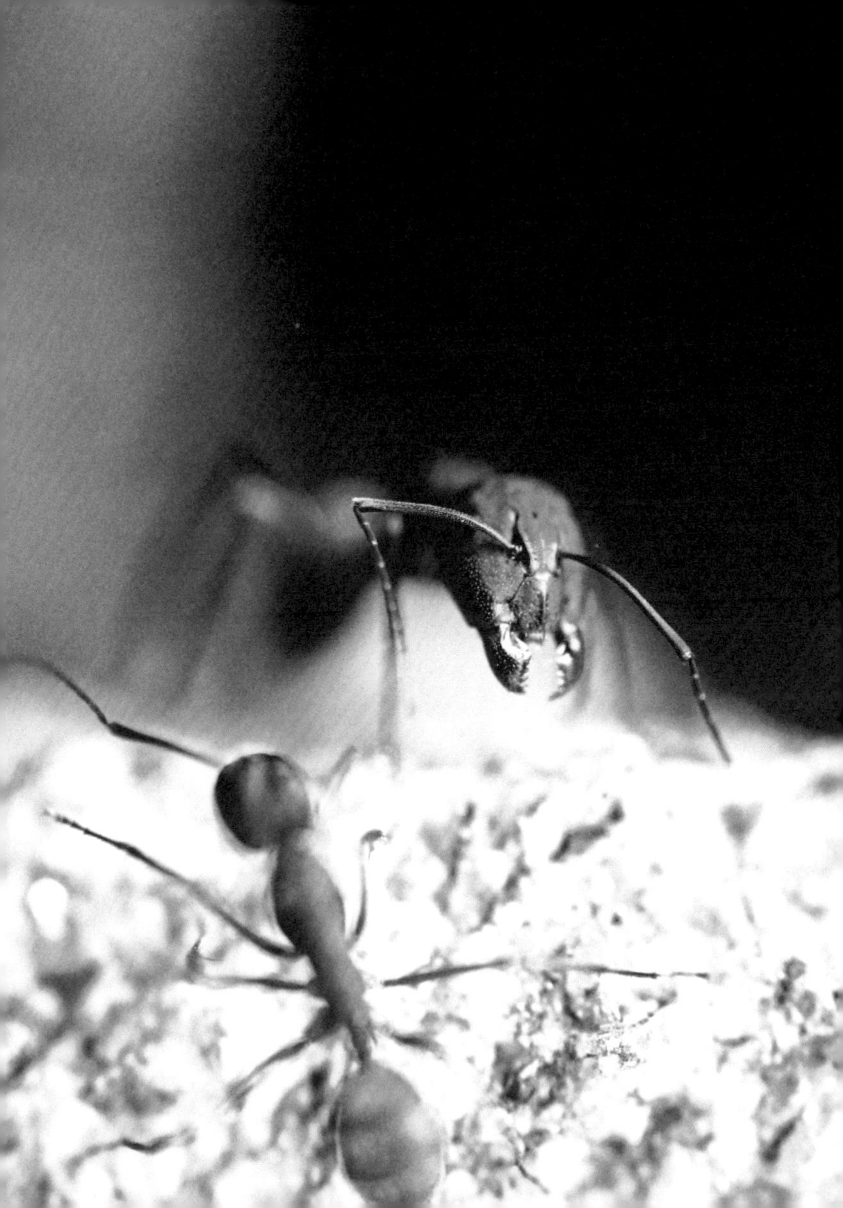

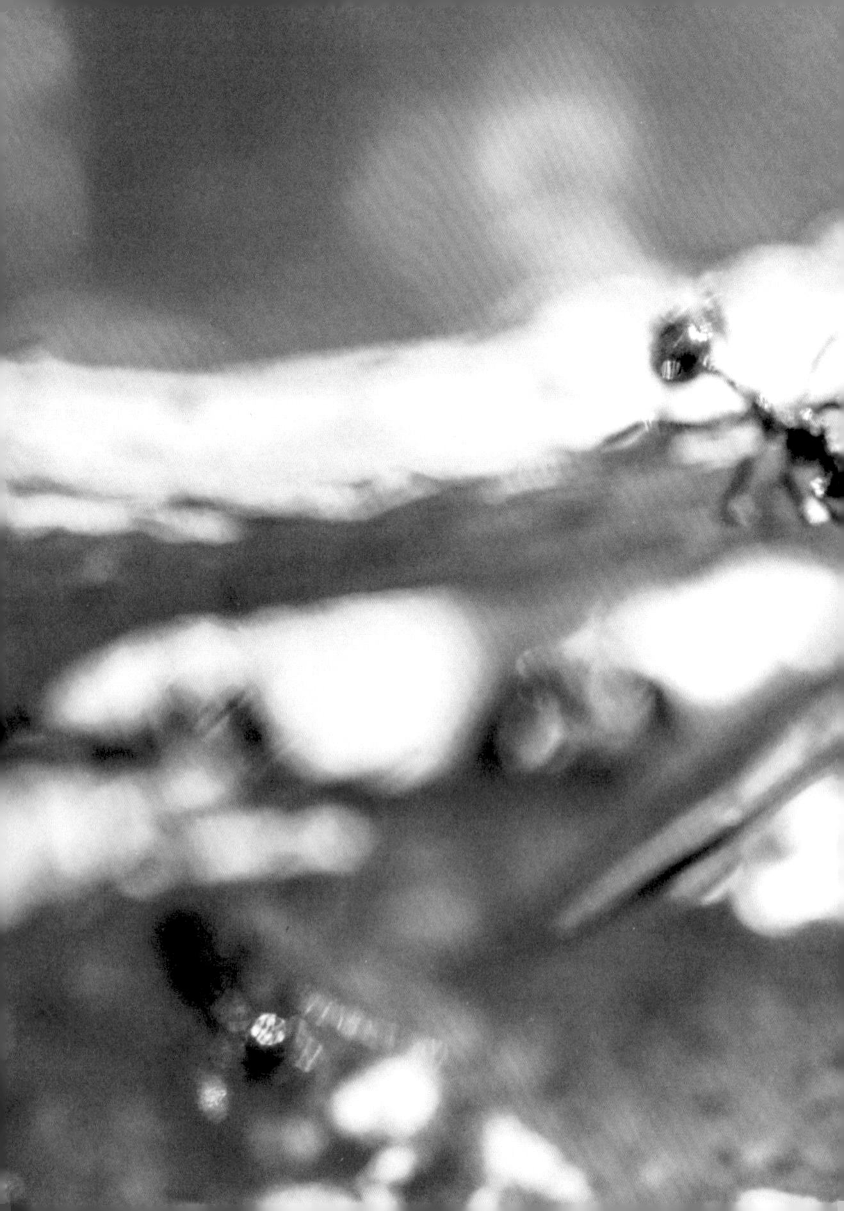

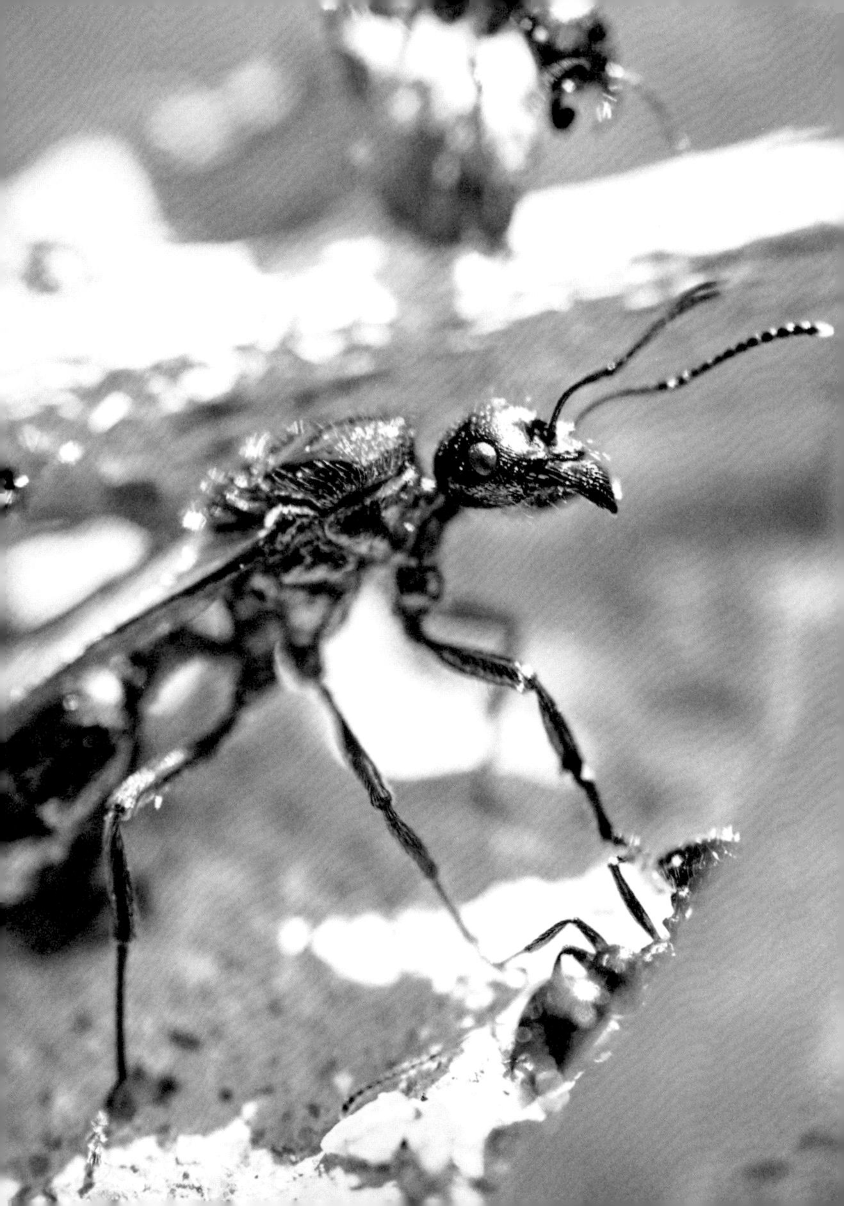

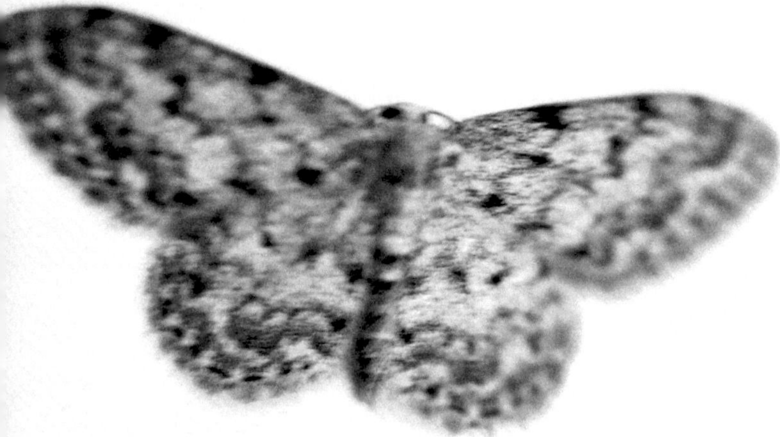

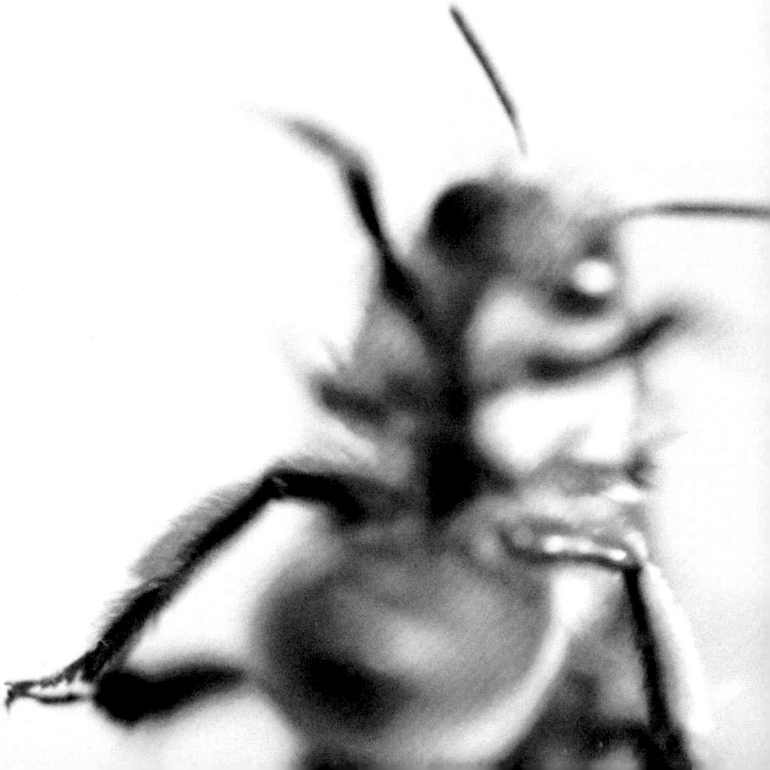

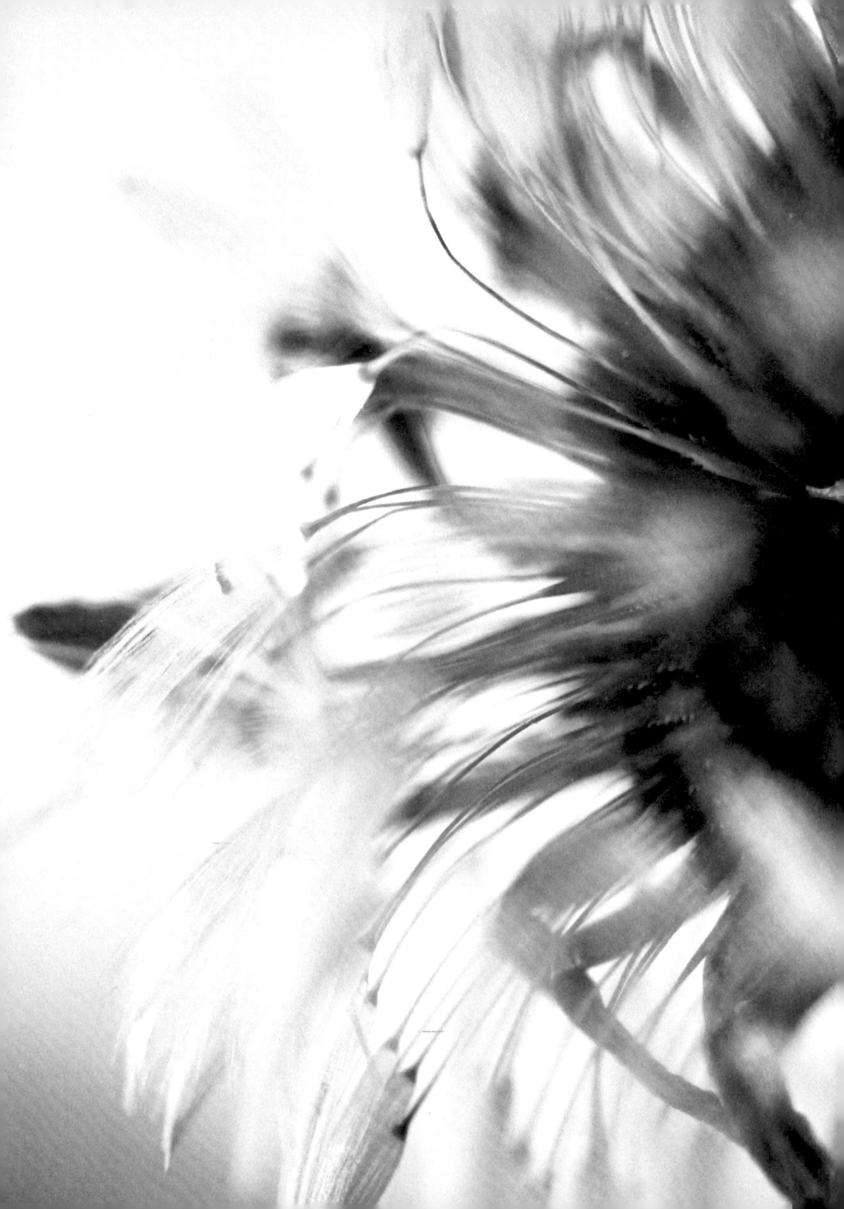

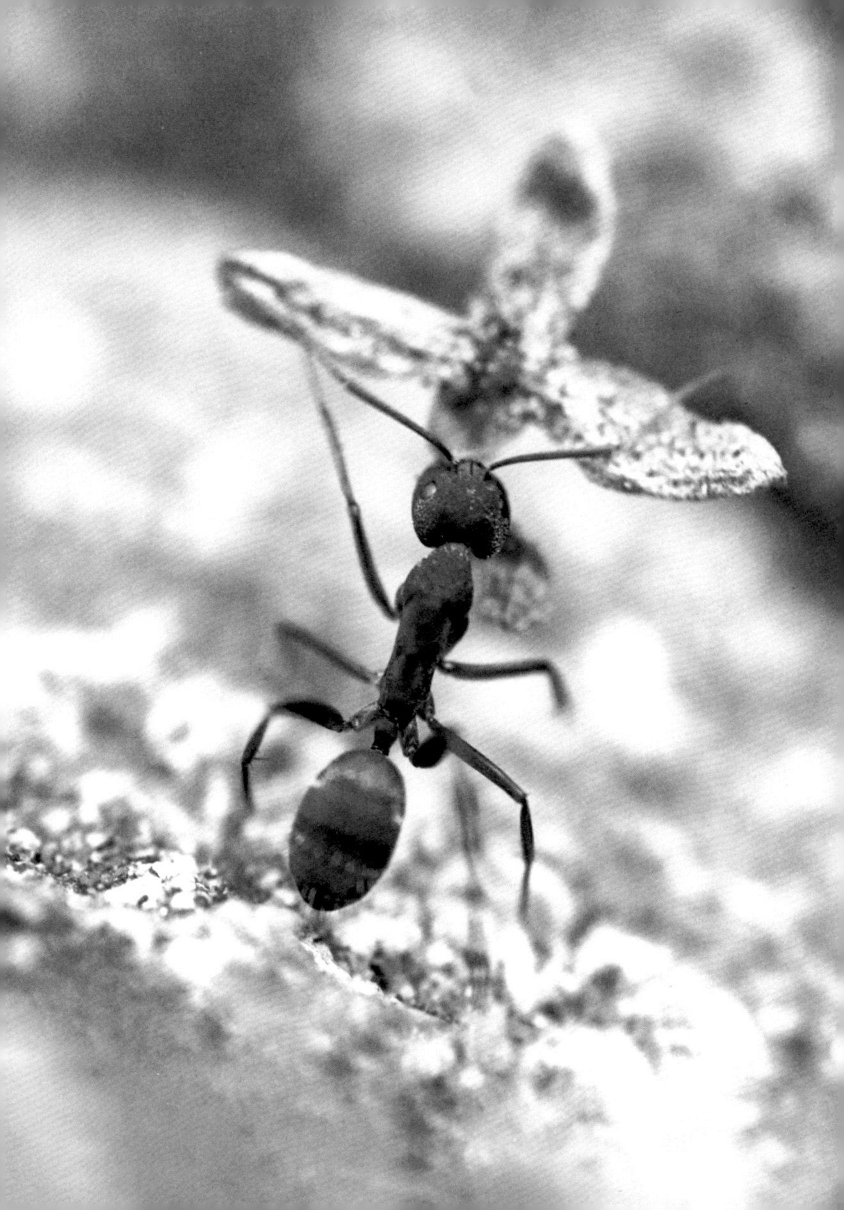

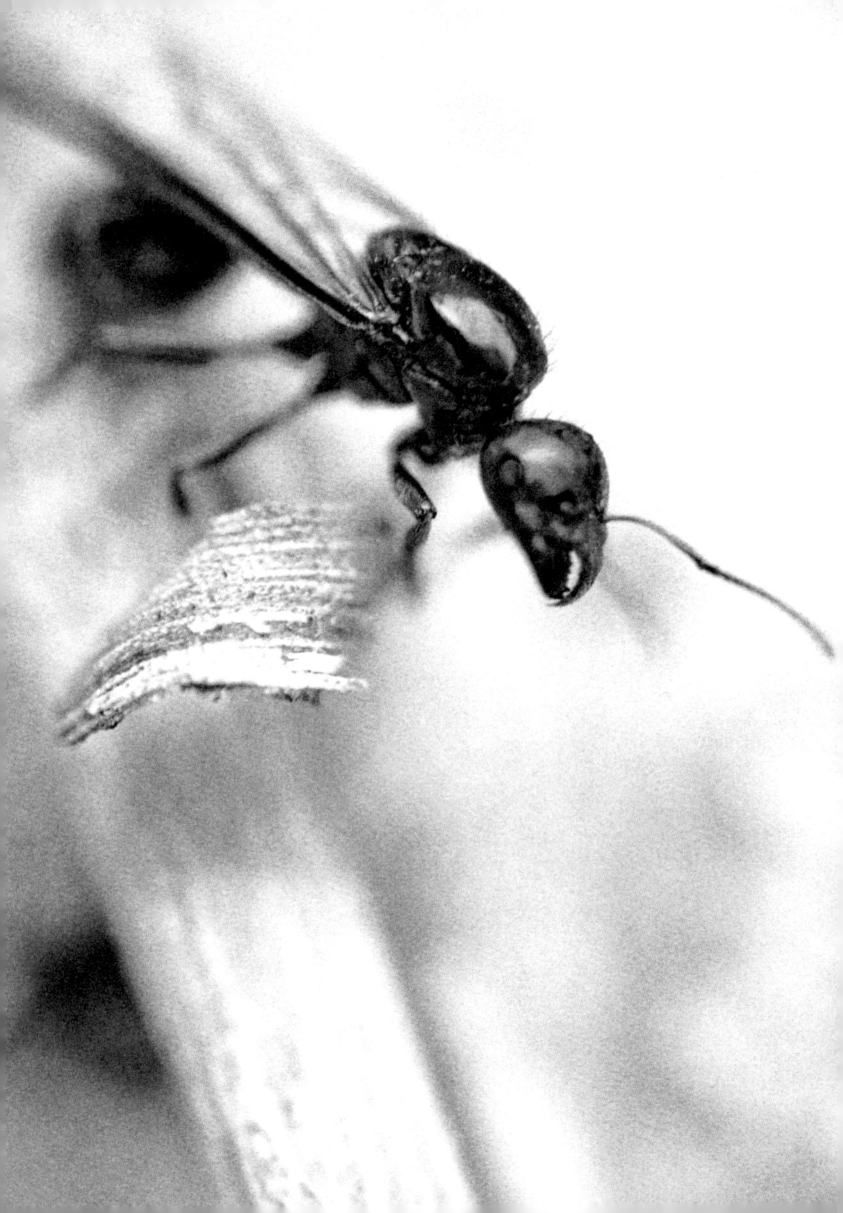

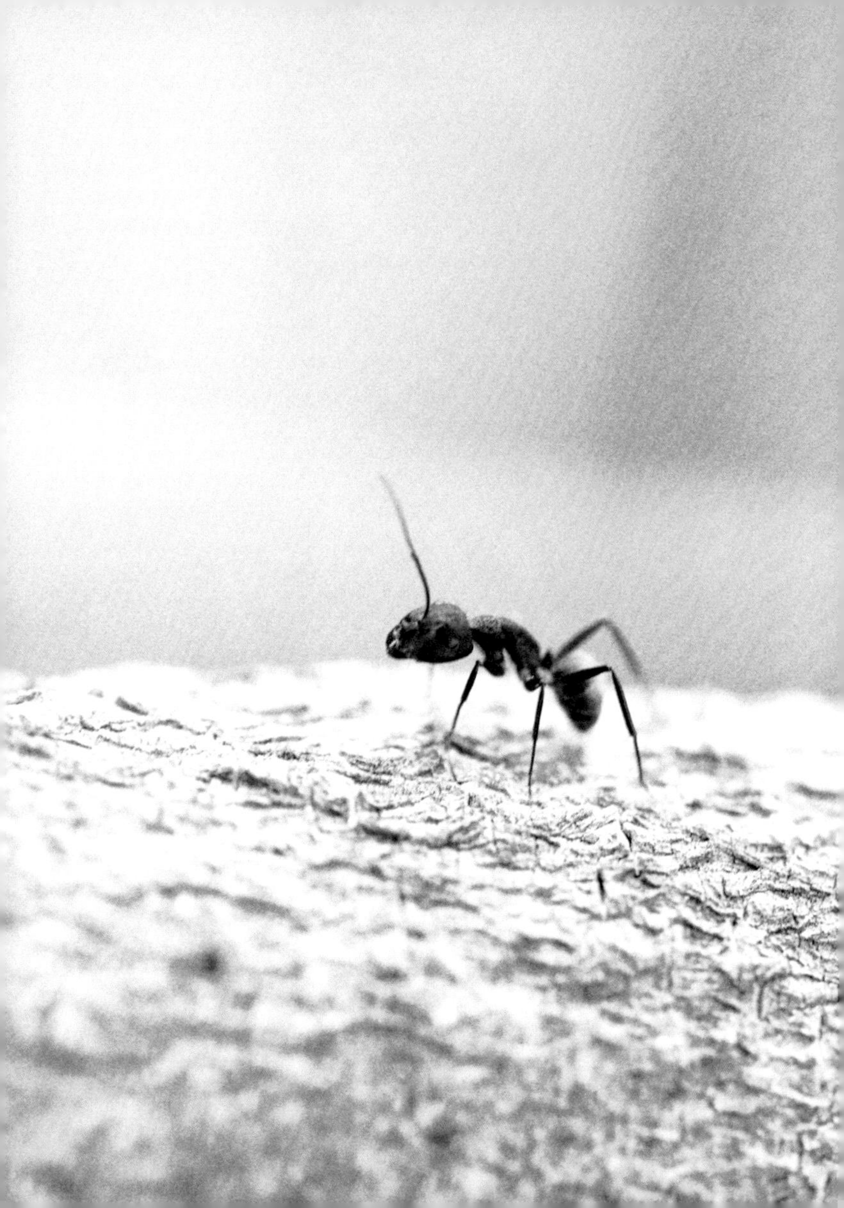

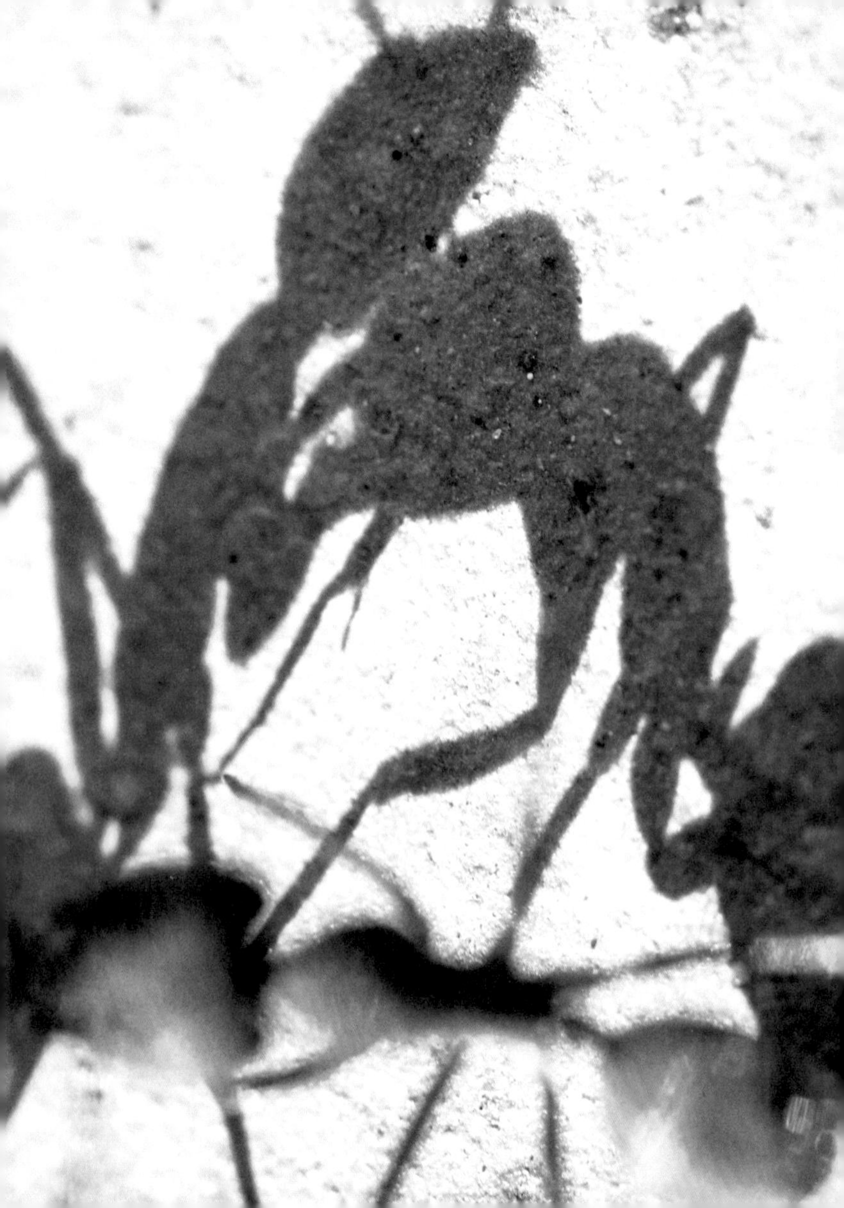

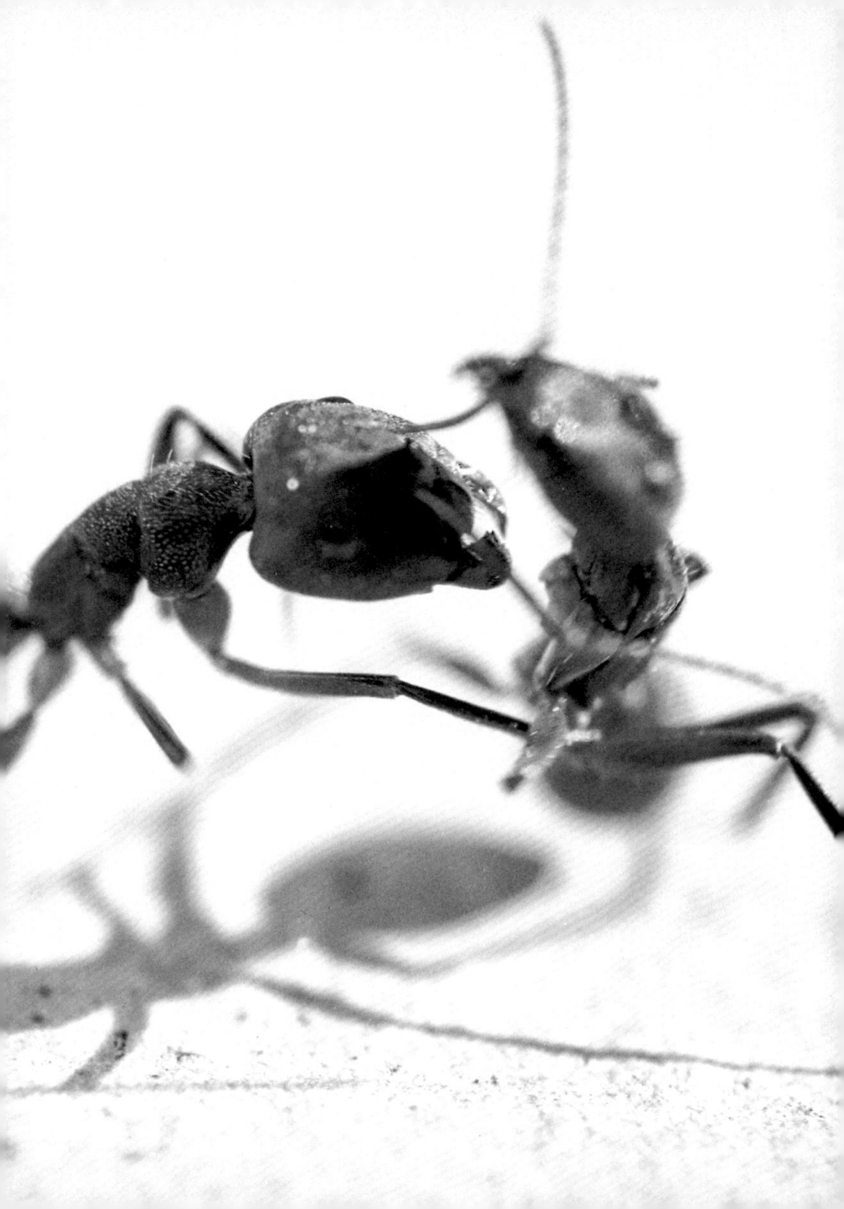

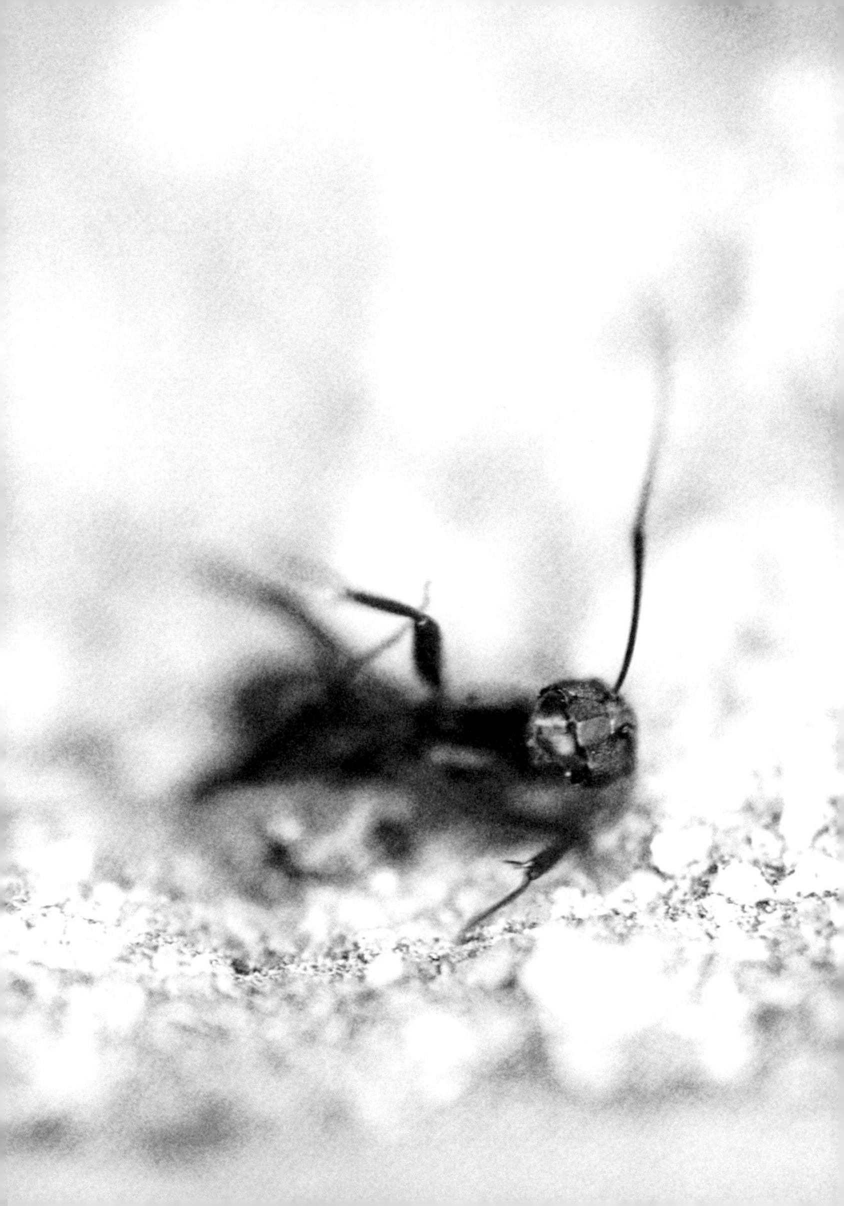

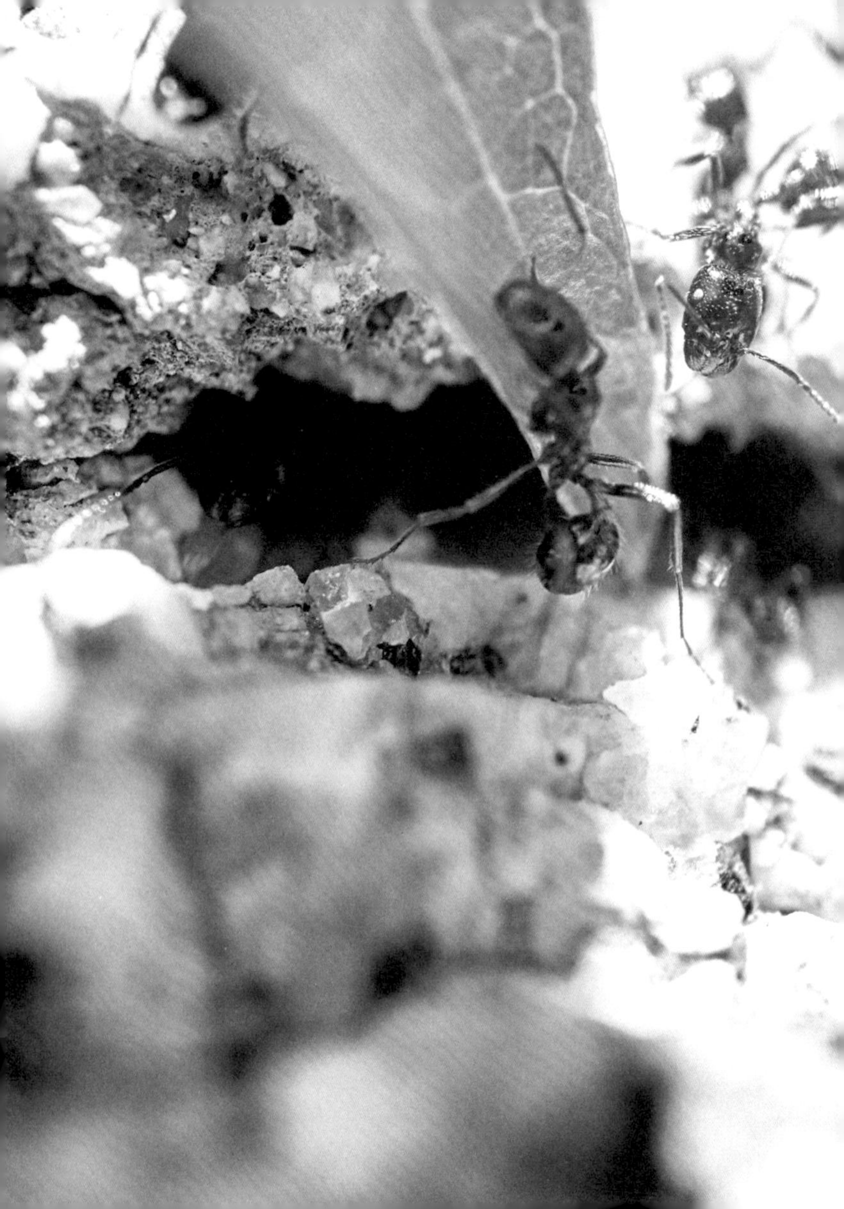

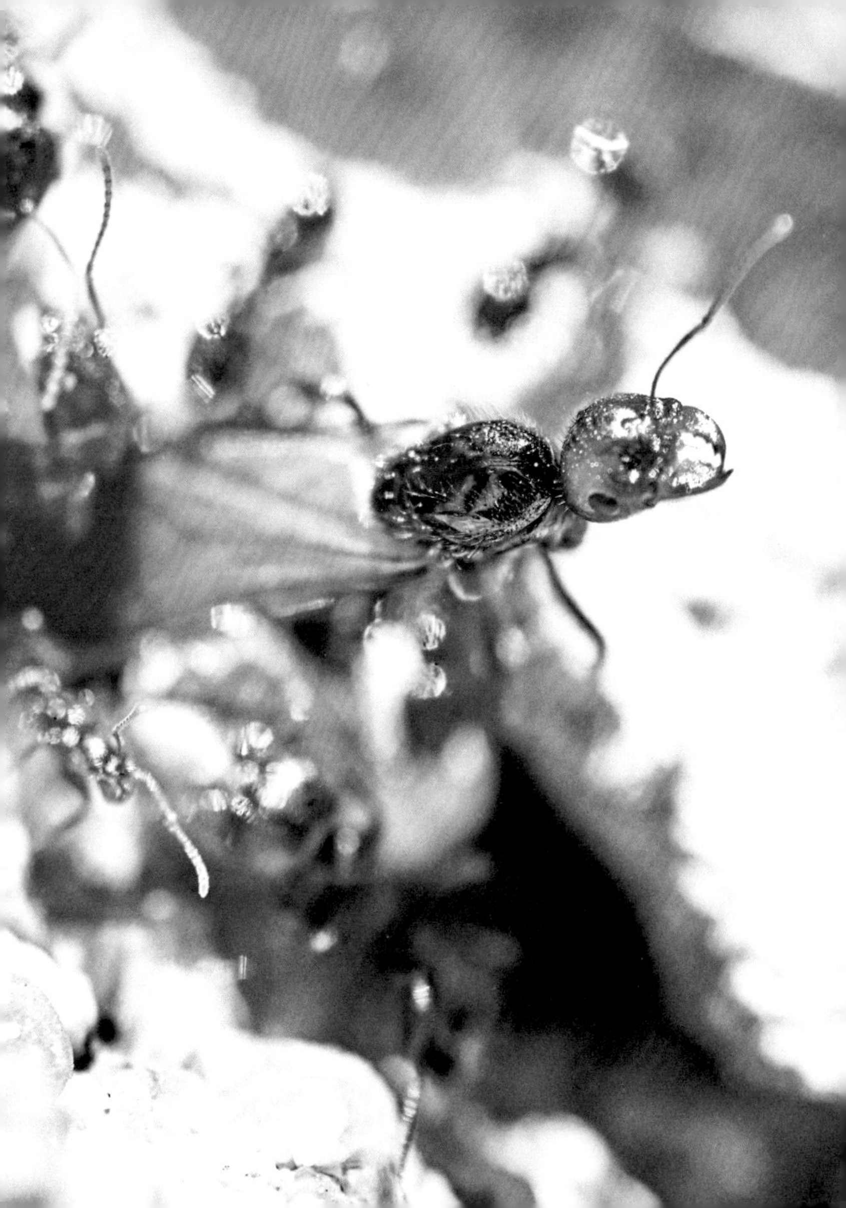

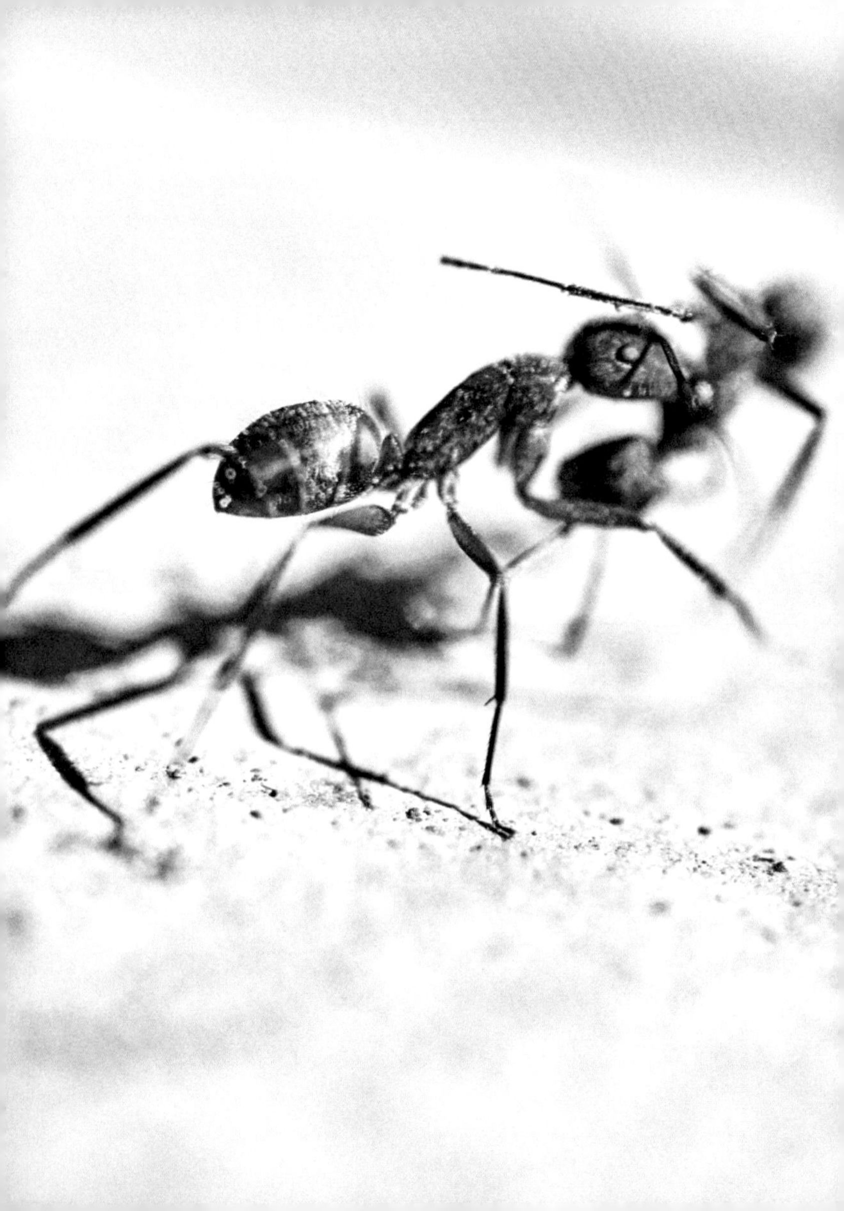

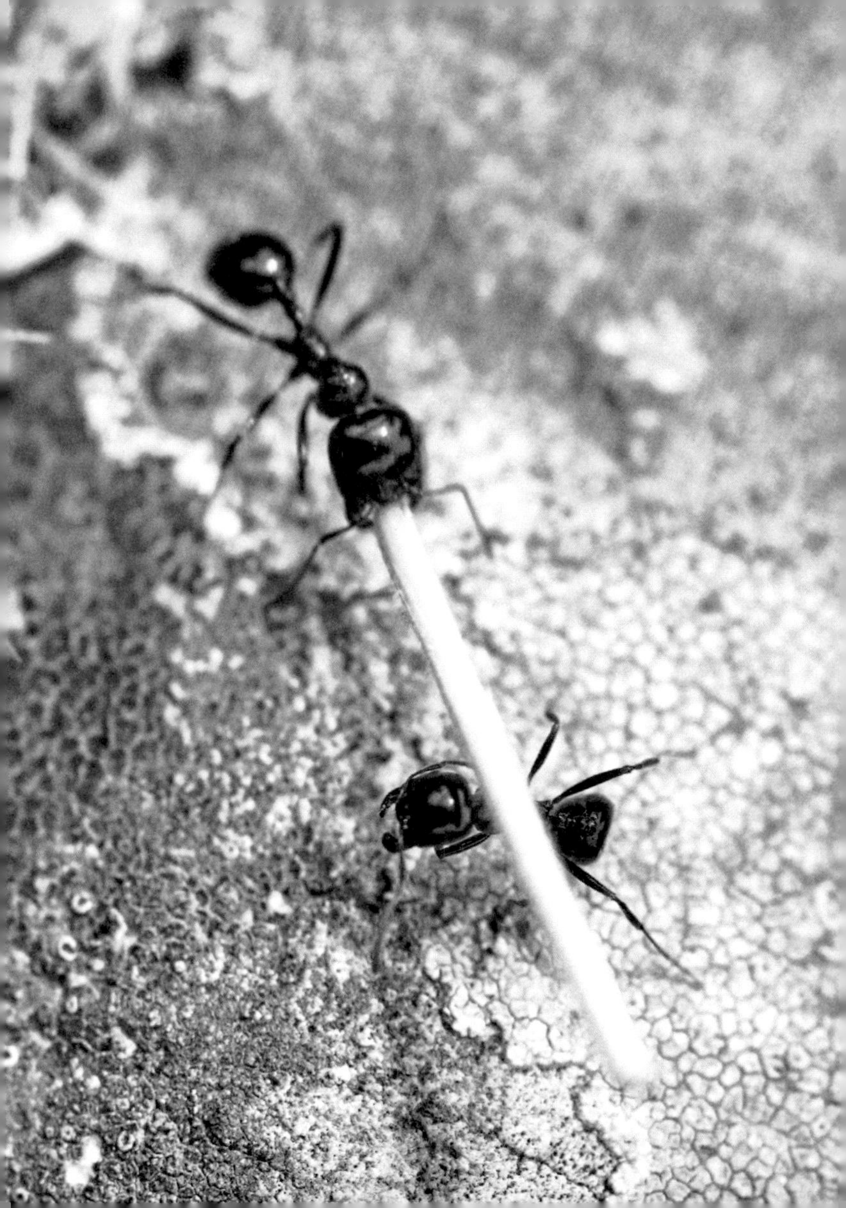

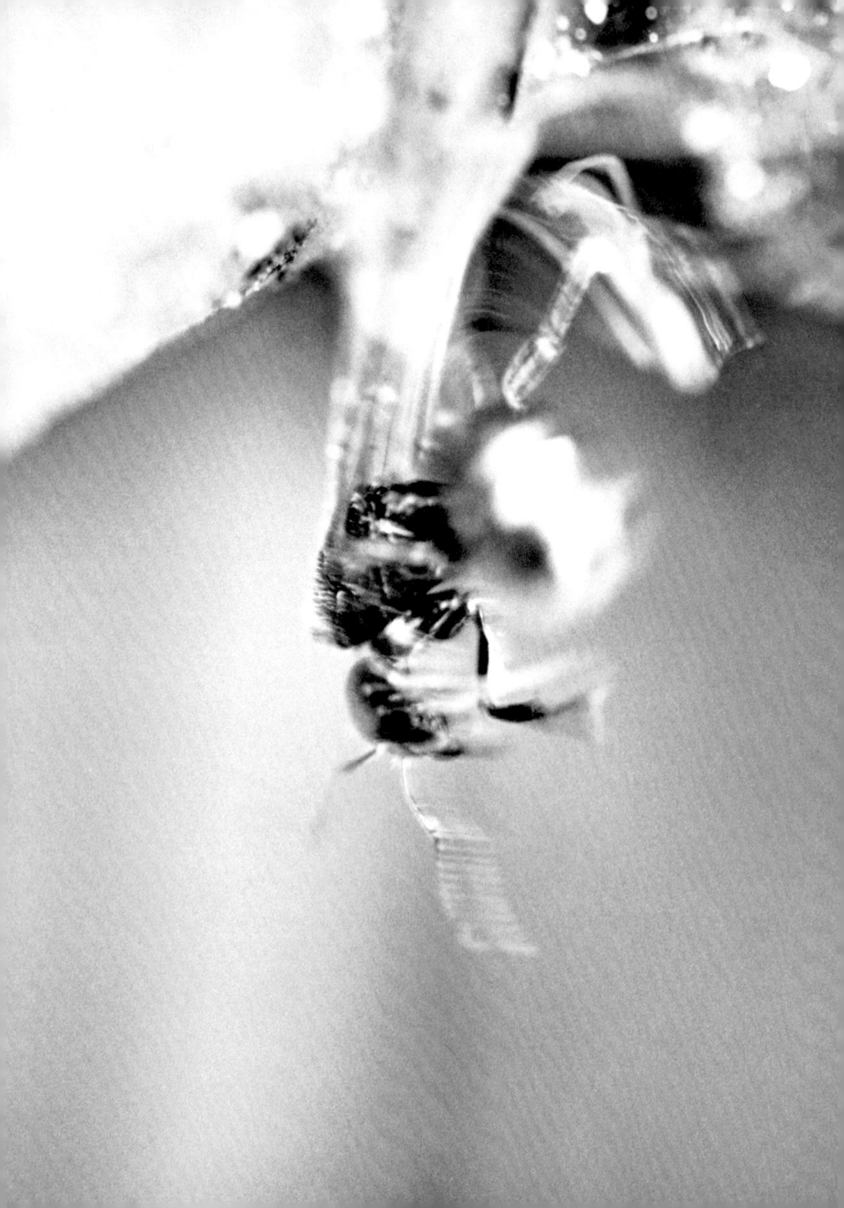

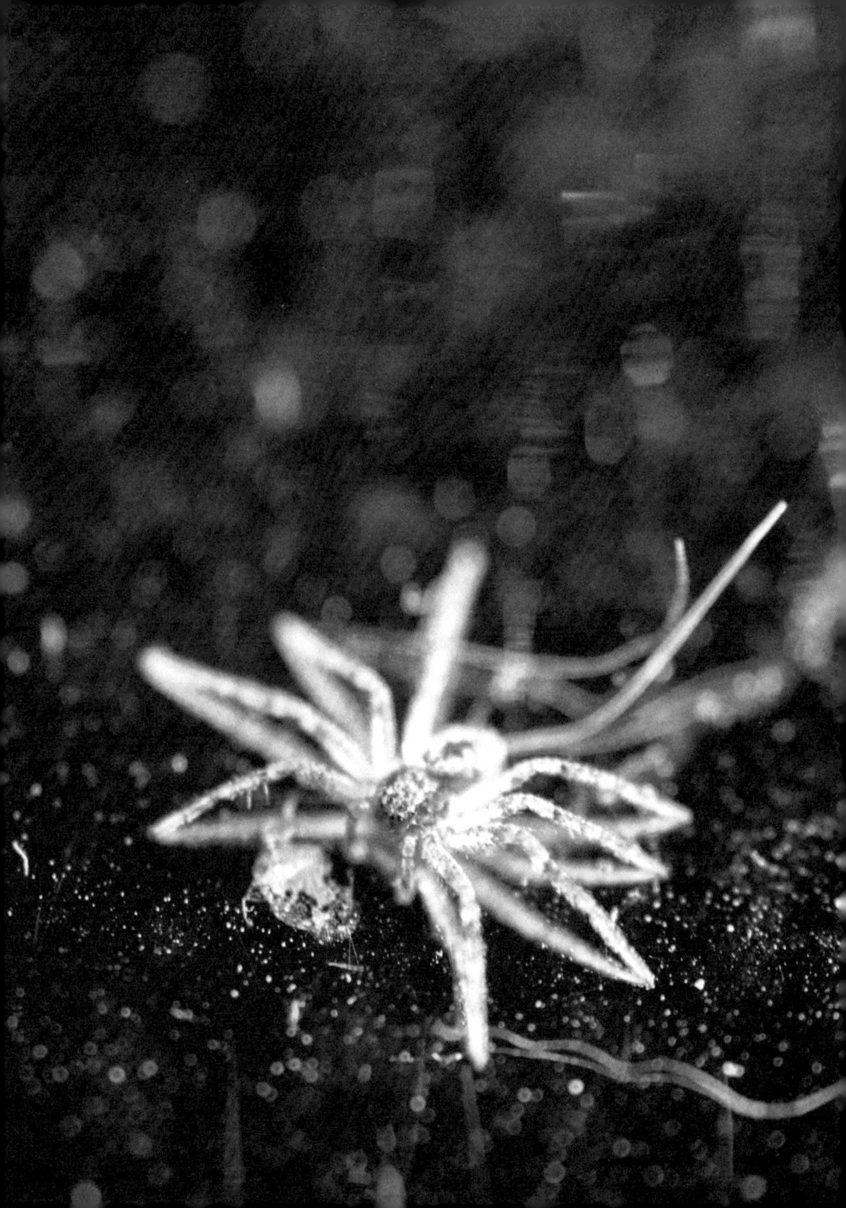

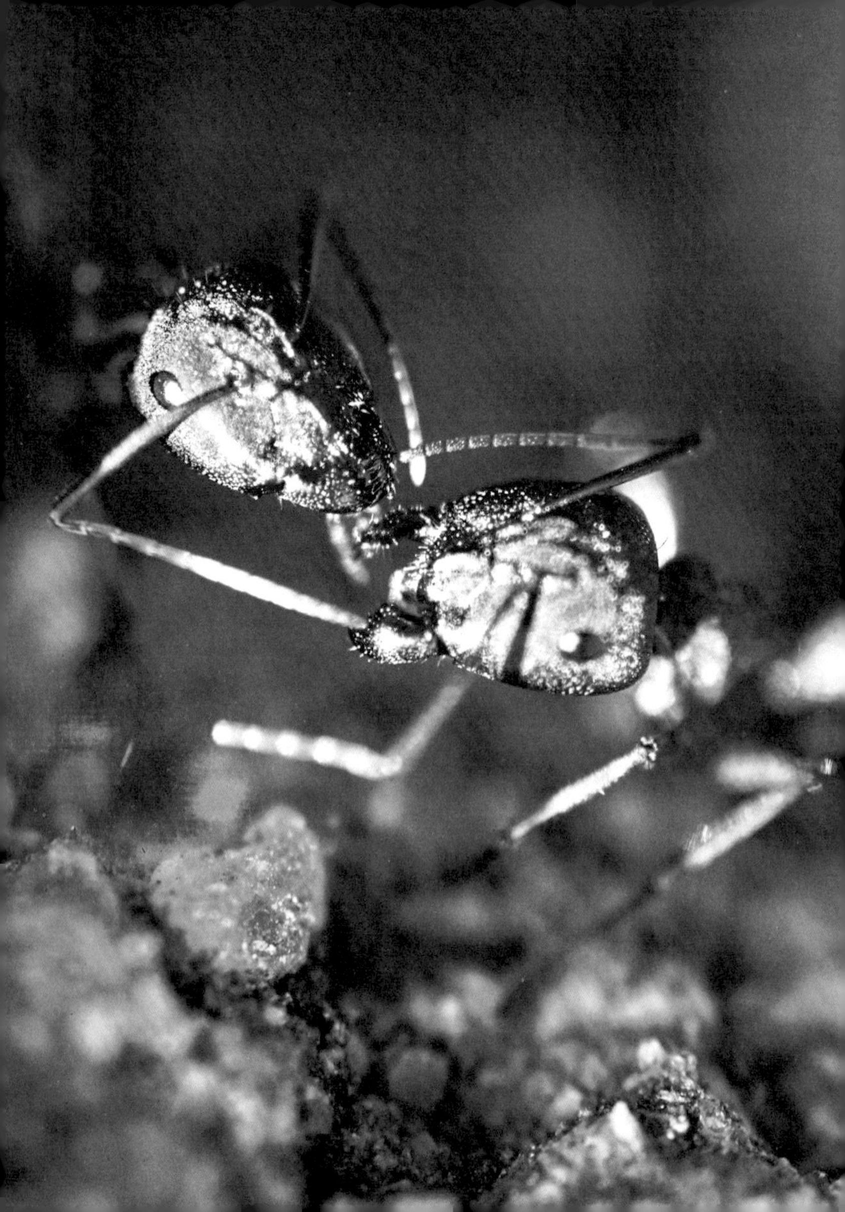

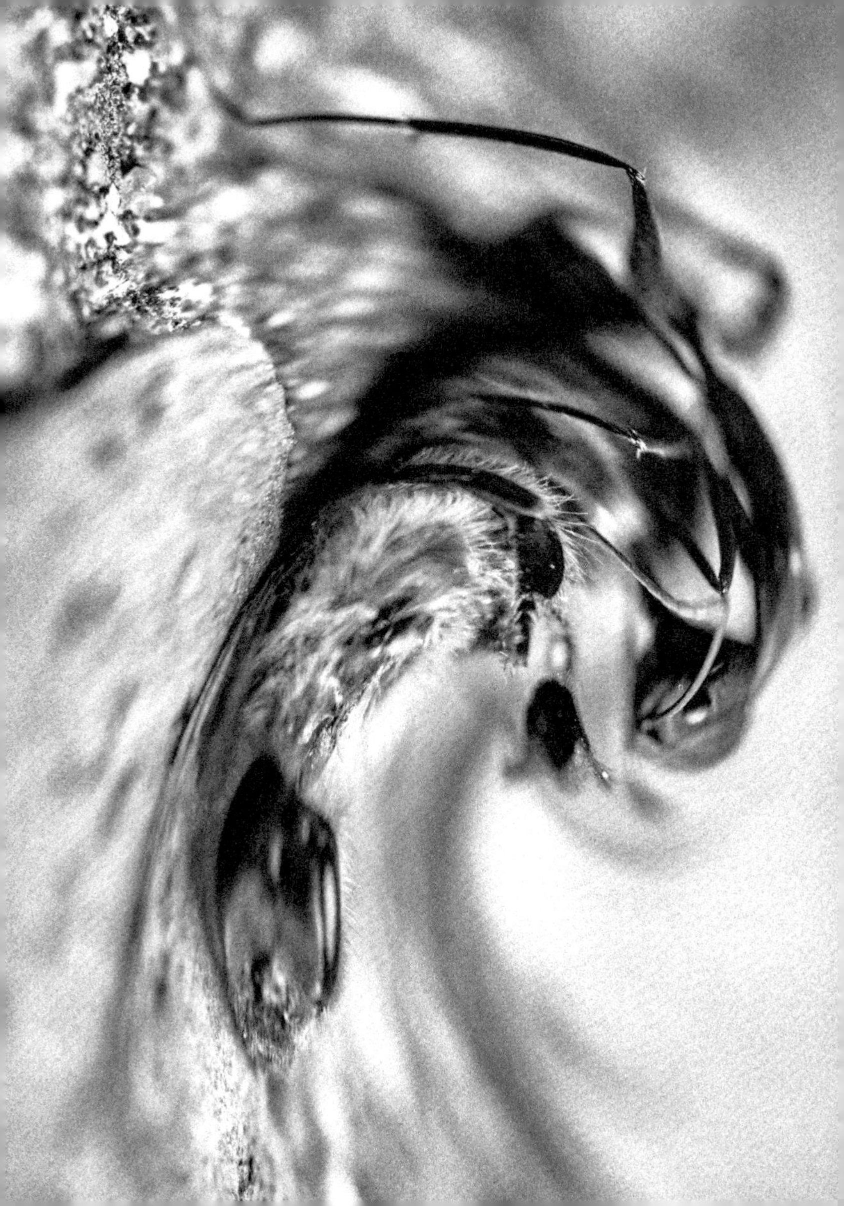

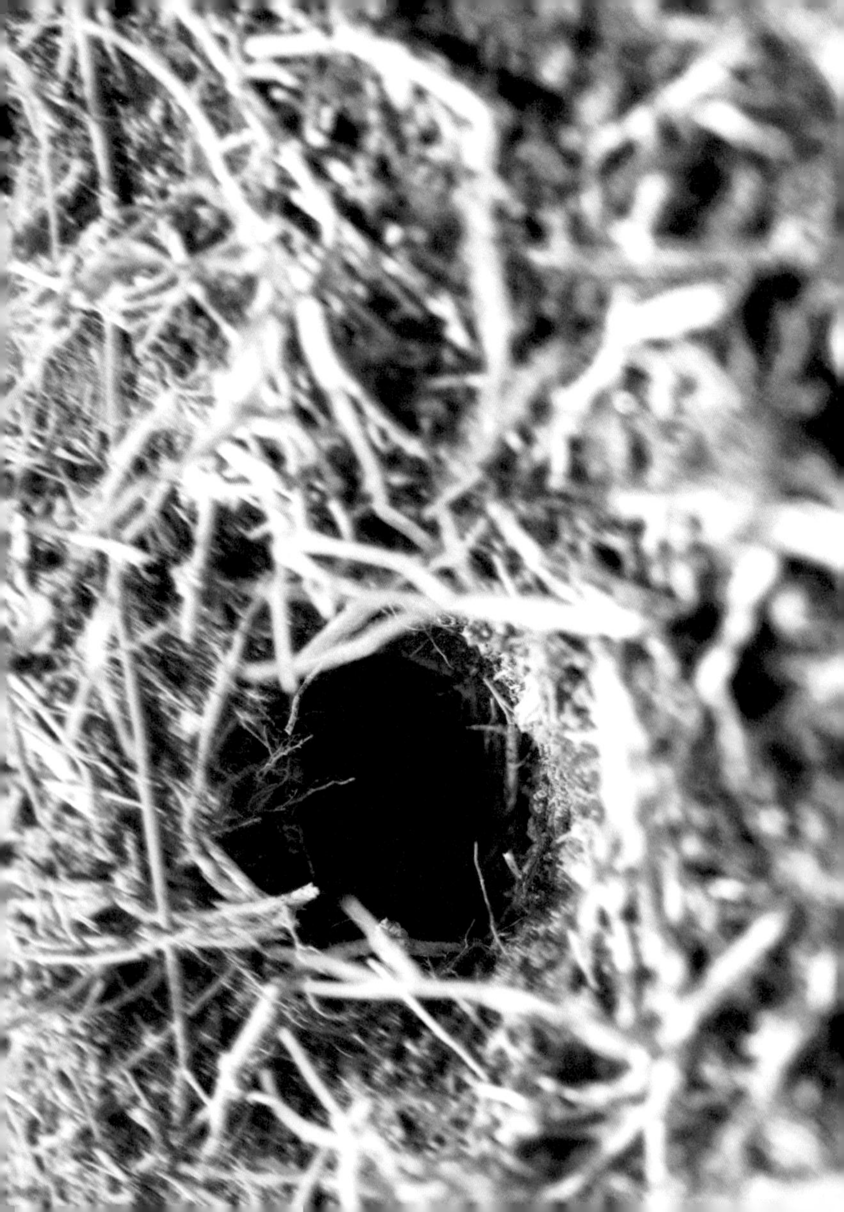

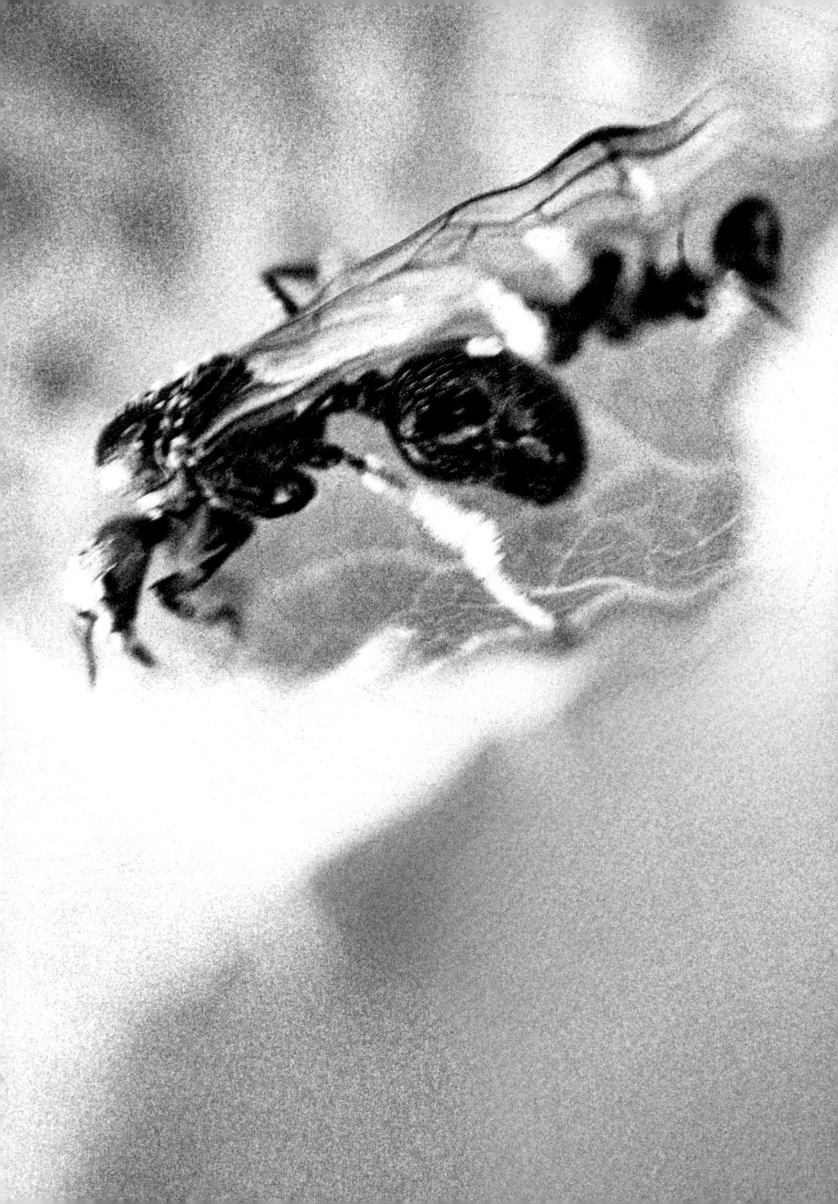

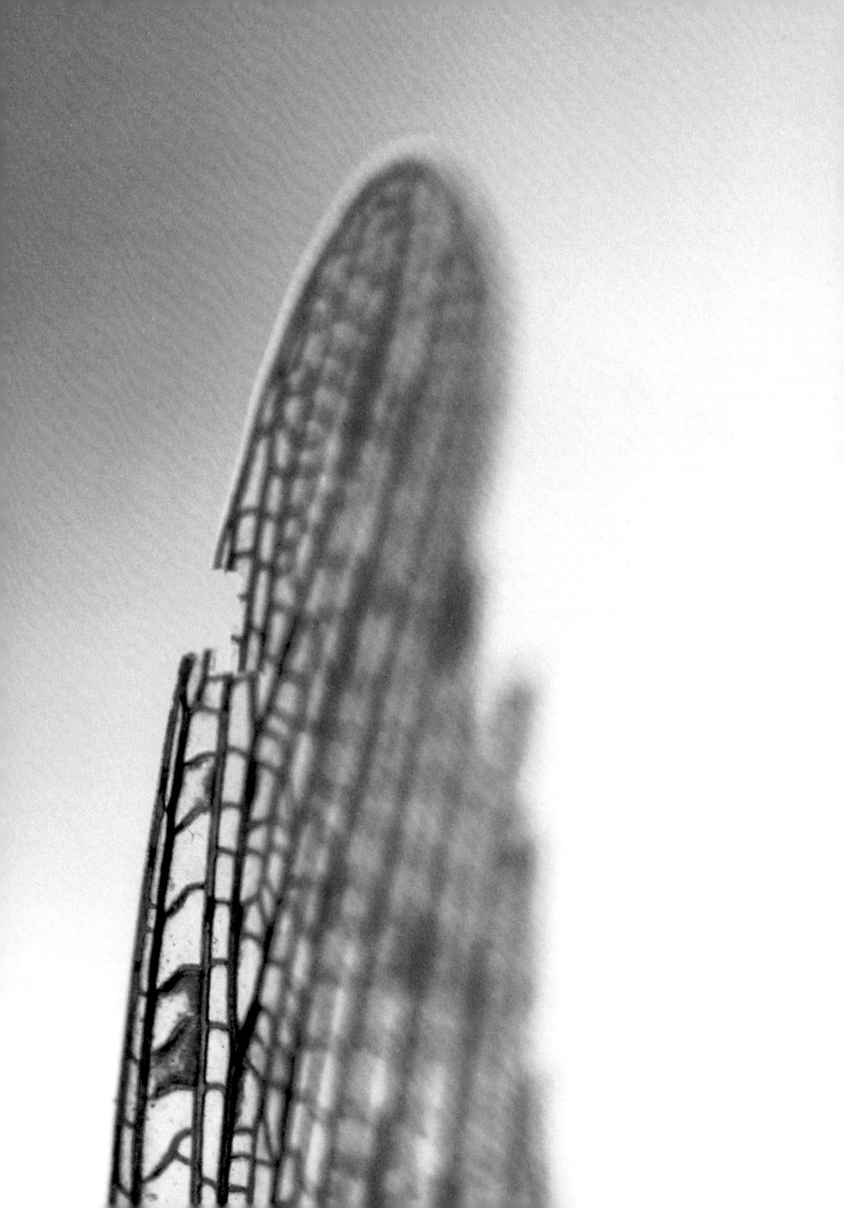

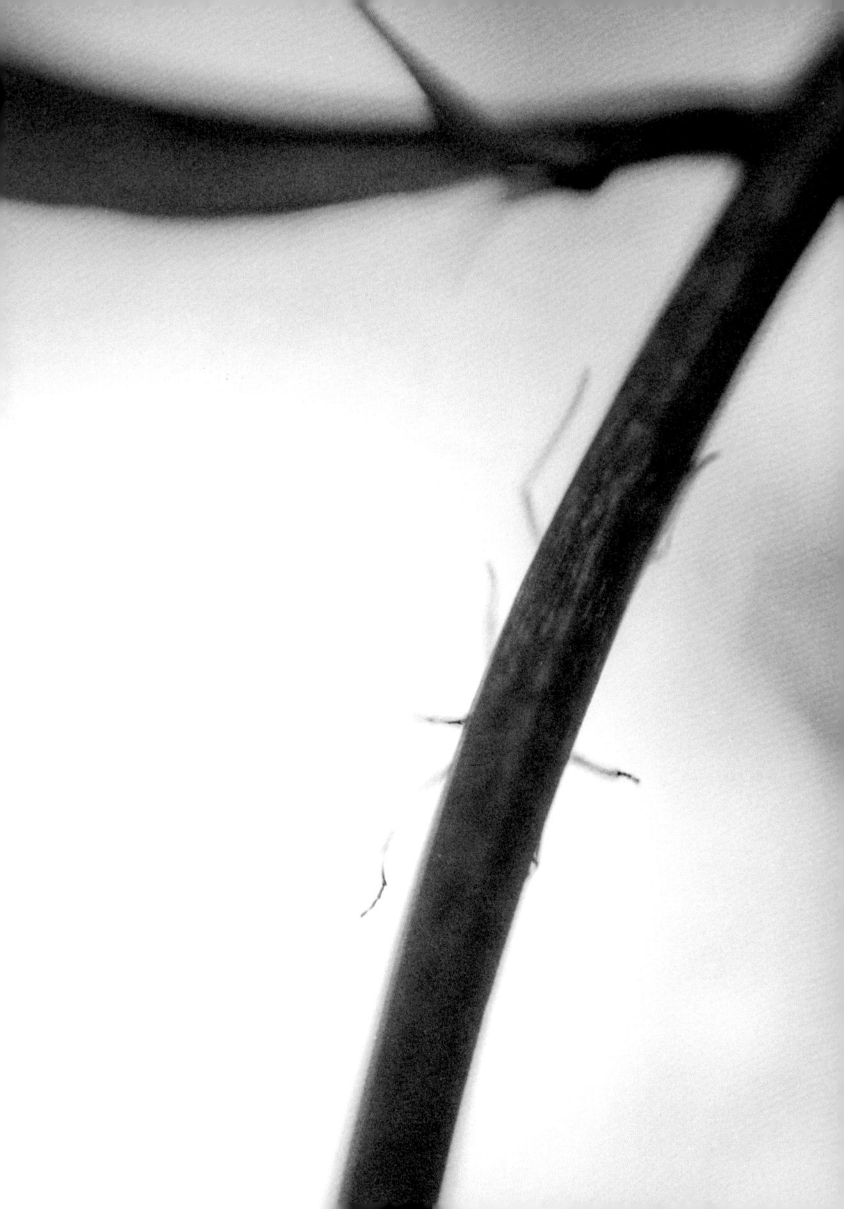

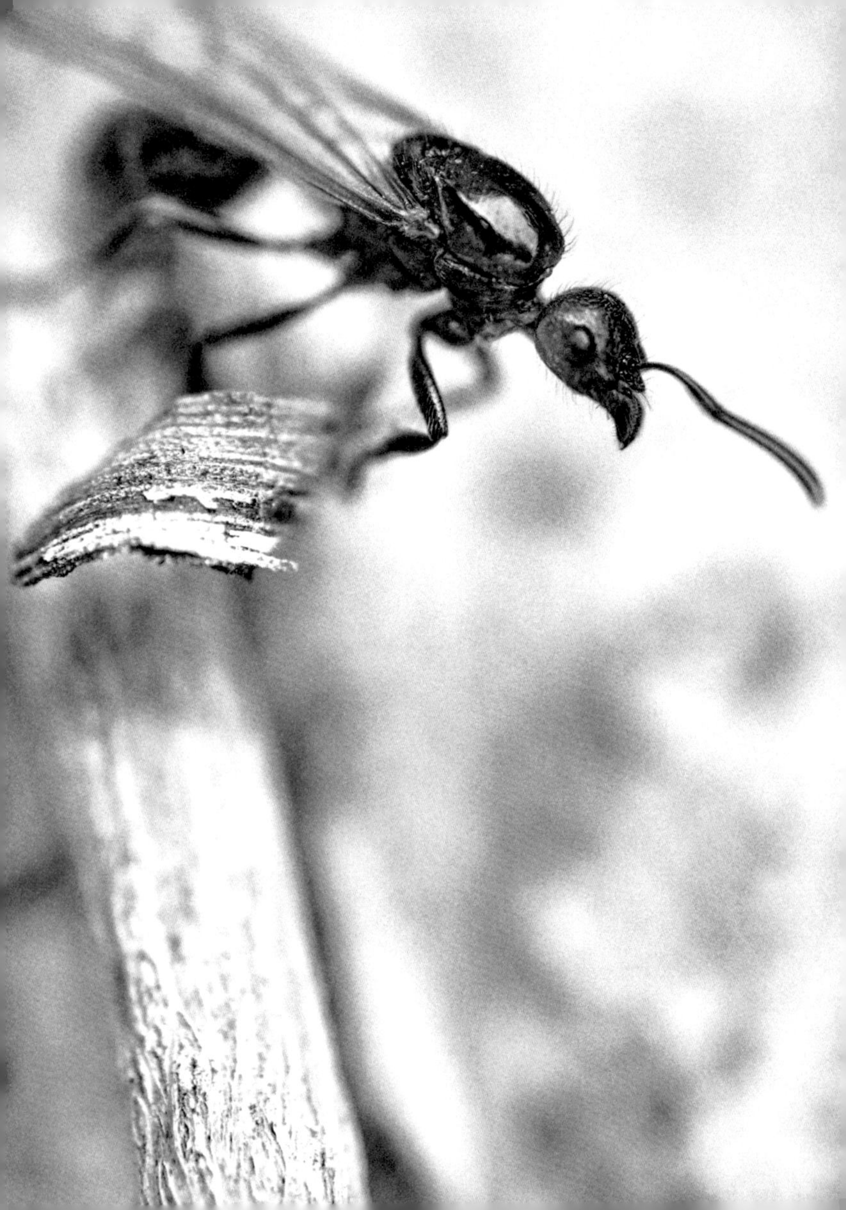

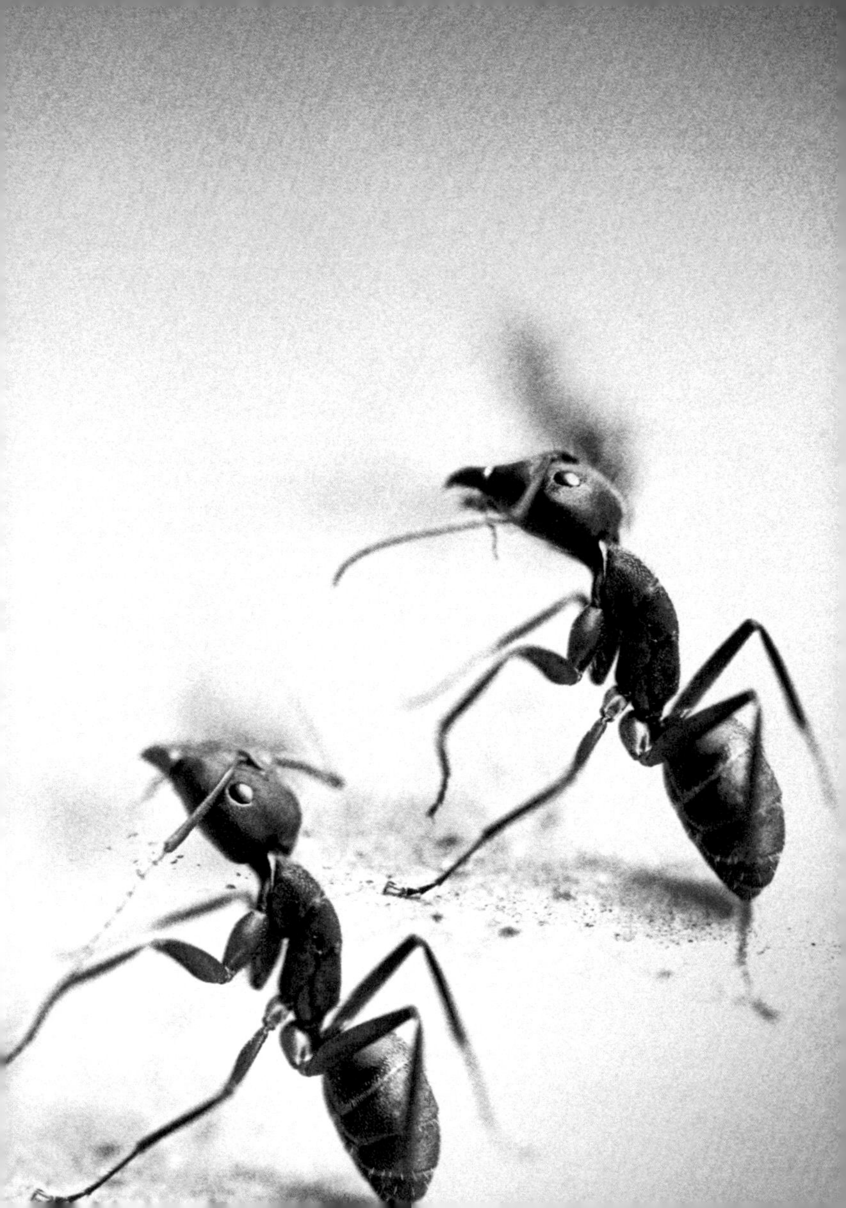

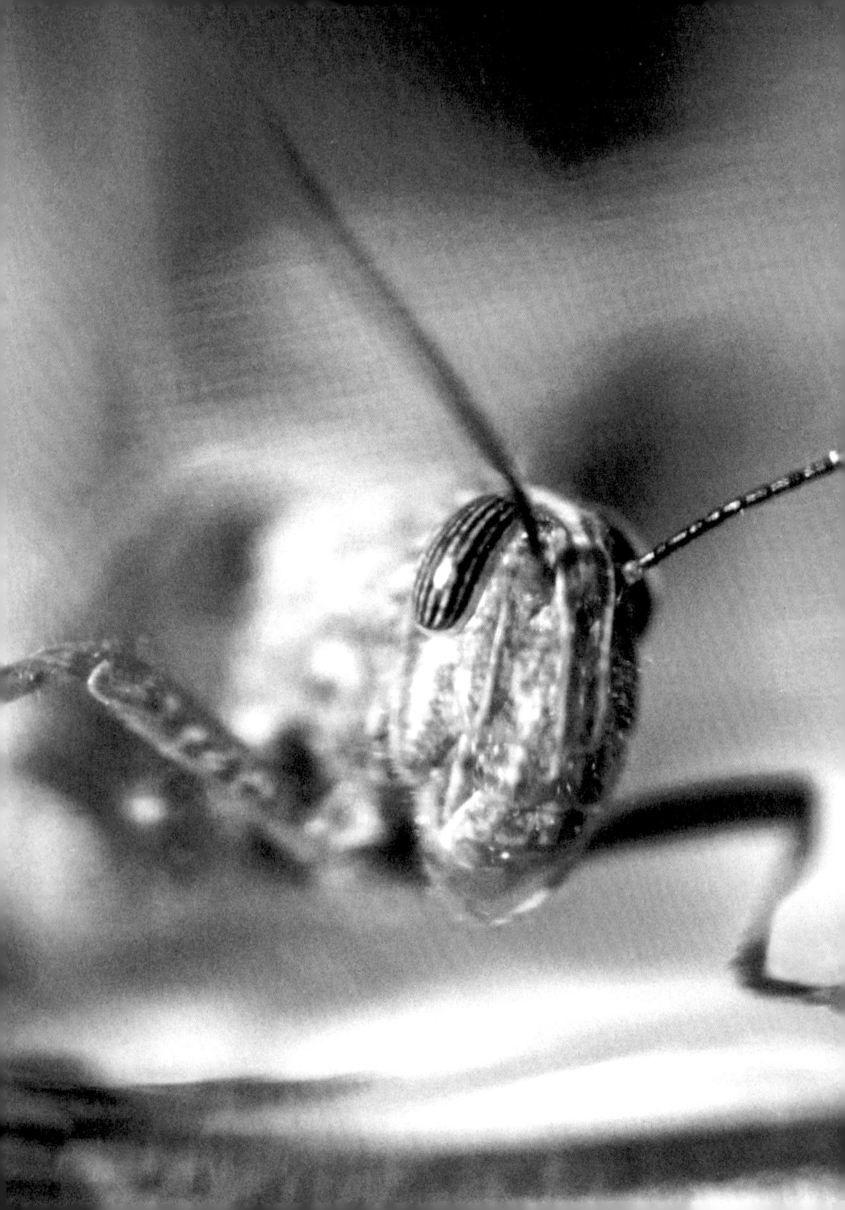

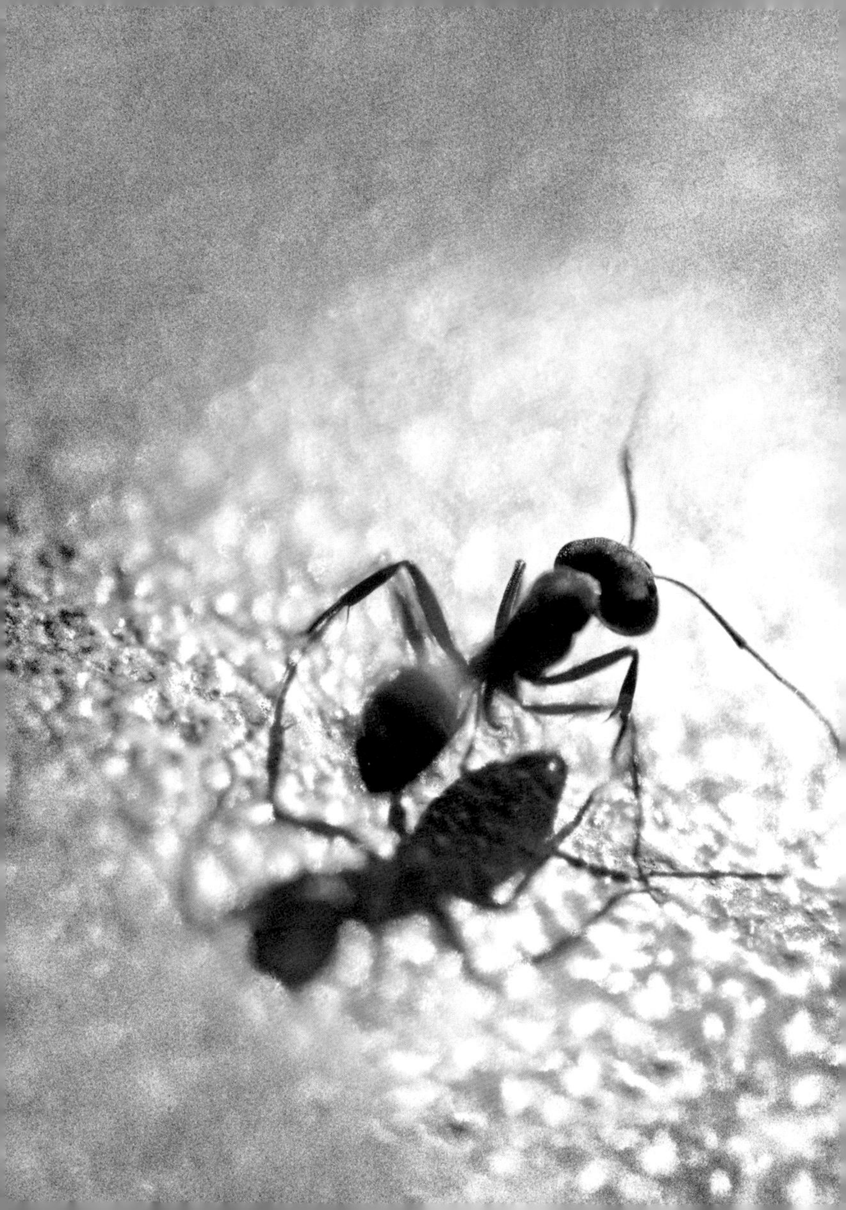

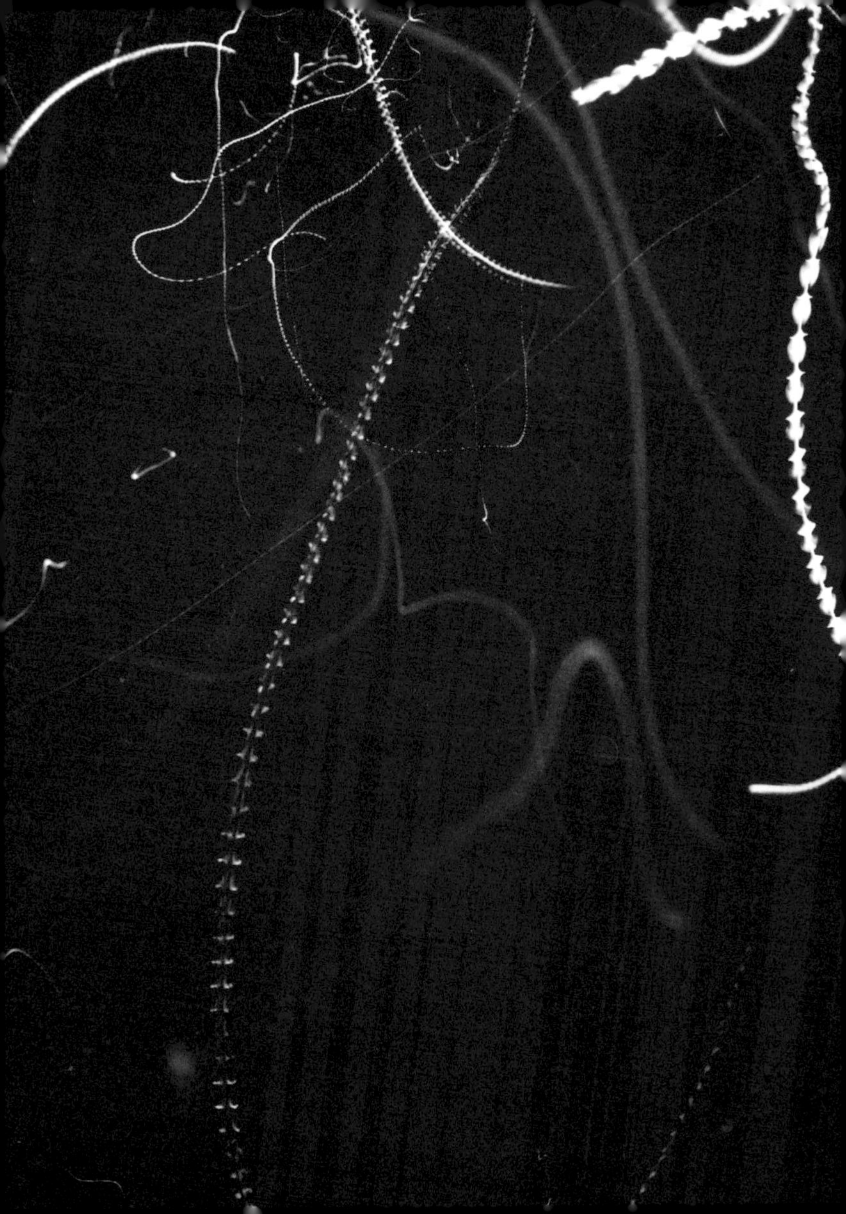

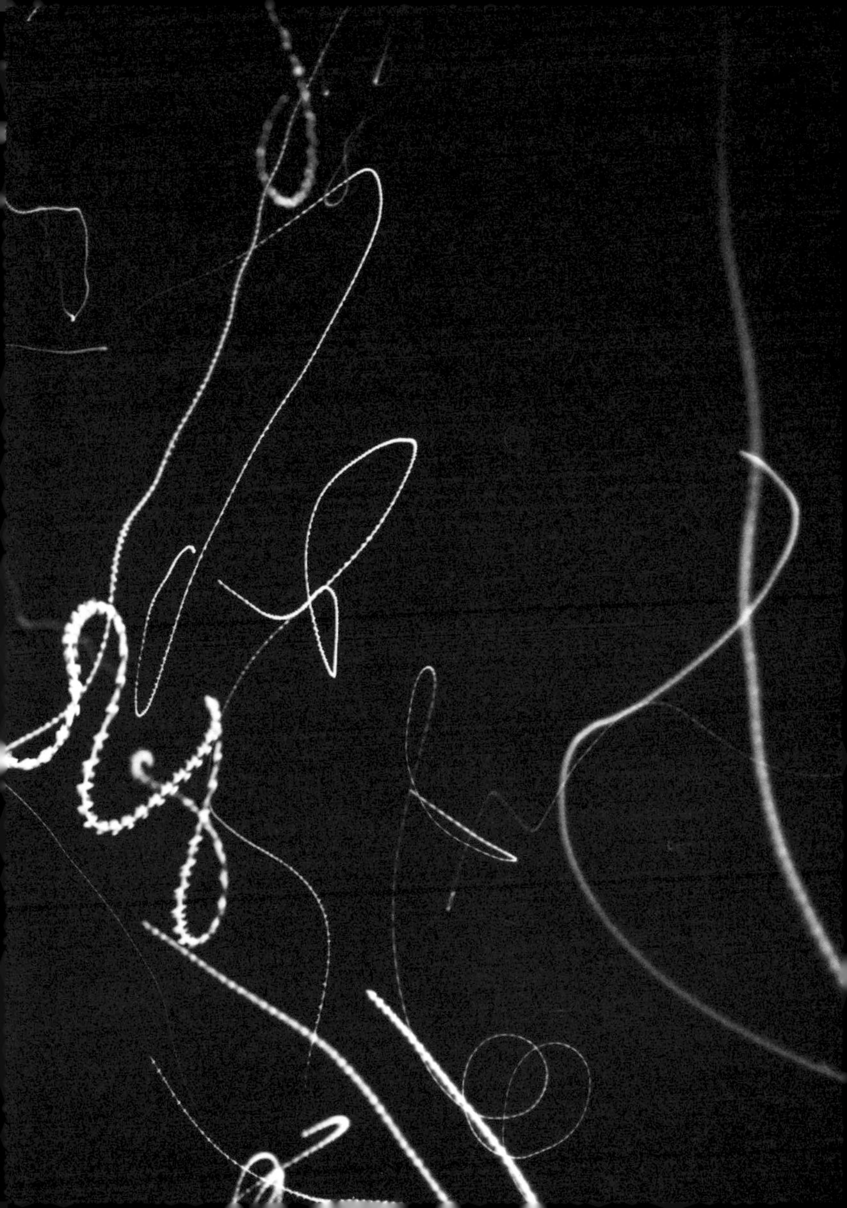

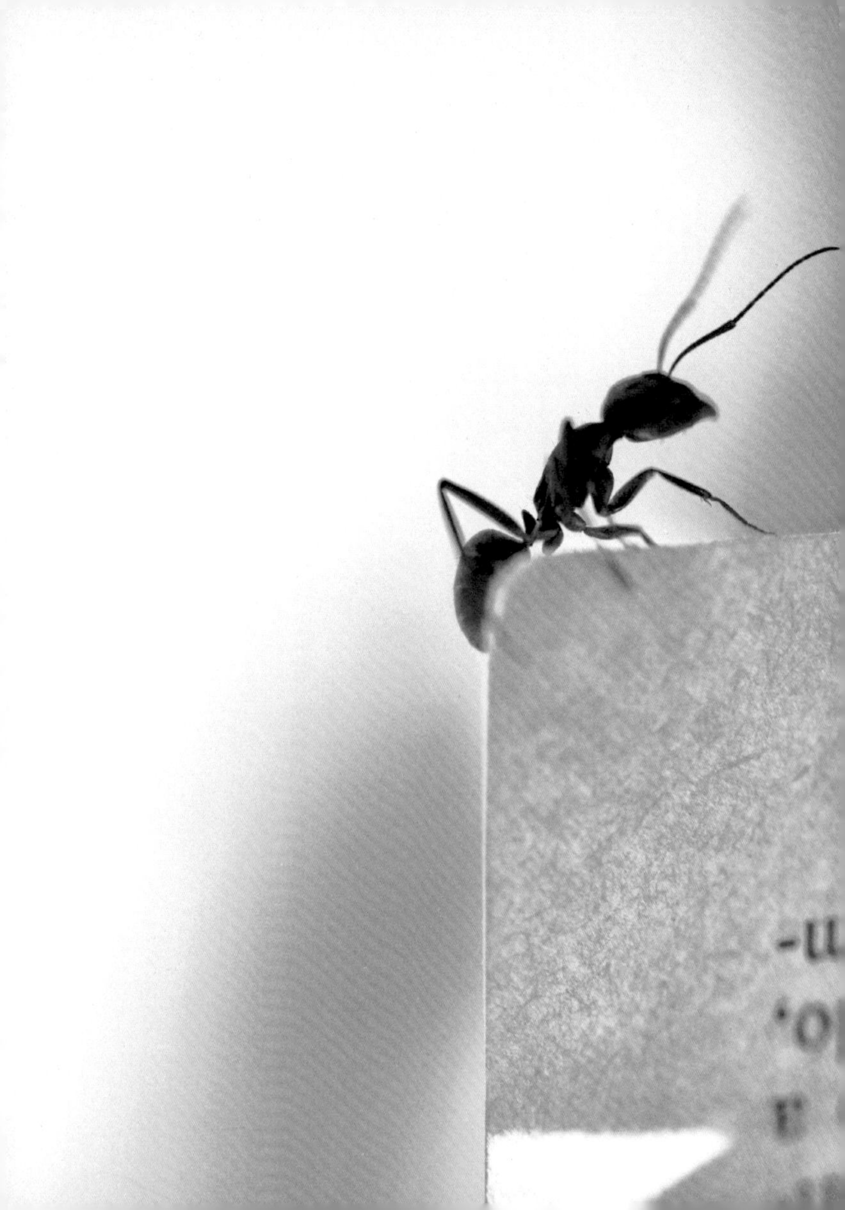